DON'T HATE
THE F WORD

THE MIKE METHOD—
LEADING THROUGH
CHANGE & CHAOS

*Stay Focused
&
Keep Leading!*

RICHARD E. MIKE

outskirts
press

DON'T HATE THE F WORD
THE MIKE METHOD - LEADING THROUGH CHANGE & CHAOS
All Rights Reserved.
Copyright © 2022 Richard E. Mike
v3.0

The opinions expressed in this manuscript are solely the opinions of the author and do not represent the opinions or thoughts of the publisher. The author has represented and warranted full ownership and/or legal right to publish all the materials in this book.

This book may not be reproduced, transmitted, or stored in whole or in part by any means, including graphic, electronic, or mechanical without the express written consent of the publisher except in the case of brief quotations embodied in critical articles and reviews.

Outskirts Press, Inc.
http://www.outskirtspress.com

ISBN: 978-1-9772-4662-2

Cover Photo © 2022 www.gettyimages.com. All rights reserved - used with permission.

Outskirts Press and the "OP" logo are trademarks belonging to Outskirts Press, Inc.

PRINTED IN THE UNITED STATES OF AMERICA

TABLE OF CONTENTS

Preface i
Introduction v
Don't Hate the F Word Pre-Thought x
1. Foundation – Is the ground soft enough to
   lead change? 1
2. Find it! Find the Talent in You. Find the Talent in your
   team 21
3. Fundamentals – Some things You Must Get
   Right Every Time 28
4. Fun – Lead with a Friday DNA 46
Prelude into the Next Set of Principles 55
5. Flexibility – Leaders can't wear the same
   pants every day 57
6. Failure – Create a Fail to Learn environment 69
7. Facilitate – Good leaders facilitate their offense! 79
8. Futuristic – Innovative thinking keeps you
   in the game! 87
9. Focus – Use your GPS to get to your destination 97
10. Fit – Are you Fit to Lead? 113
11. Fearless – Turn your hat backwards and Jump in! ... 122
12. Follow Through – No Excuses. Get it Done! 134
A Few Shout out's! 140
Acknowledgement 142

PREFACE

Meet the Author Richard Mike as he shares his "Childhood Breakthrough Moment" Defining his walk in Leadership

There I was, a 16-year-old junior in high school, on a Saturday afternoon in May 1992, standing on the starting line of Todd Stadium in Newport News, Virginia. I was among 25 other runners preparing for my biggest race of the track and field season. This was the Virginia Eastern Regional 3200 meter run, a 2-mile, 8 lap event. I was ranked third in the region and had yet to win a major race during the outdoor season. Based on my prior runs, I was projected to finish at best in third or fourth place.

Despite that projection, I had a reputation for winning. As a freshman in 1989, I was known at school as the kid with no lungs. As my teammates would say, I could run for days and never stop. But this particular season was a struggle because I felt that taking on both the 1600 meter and 3200 meter runs was really leaving me at a huge disadvantage. My body wasn't able to recover within the hour between races and the competition was better.

This was my race, I told myself. And I could silence the doubters. My only obstacle was getting out of the 1600M run. I talked to my high school coach, Mr. Spellman, and he understood my desire. He left the decision up to me. I thanked him and claimed that he could count on me to win the race!

The day of the race came and, based on the team's performance in the other events, we stood a chance of winning the overall championship. This would be the first in school history. I needed to place in the top three in both the 1600 and 3200 runs to give us a chance based on the point scoring prediction. I still opted to run the 3200M run only.

As I was standing at the starting line waiting for the gun to sound, I felt I was being selfish and letting the team down. But I had to tune out the crowd noise and forget about the second thoughts so I could focus on my race. I had such an exhilarating feeling of excitement because I was prepared to win. I never came into a race with a specific plan or a strategy. I was known for just getting on the track and taking off. For this race, I actually developed a plan. I was going to relax and stay behind Jack, the #1 runner and pace myself for seven laps. While conserving my energy, I would then take off like lightning, sprinting the entire last lap.

When the gun sounded, I took off and by the end of the first lap I positioned myself as planned in 2nd place. And for seven laps I paced myself in 2nd right behind Jack to save my energy for the last lap. But on three occasions, the 3rd, 4th, and 6th lap I was tempted to breach my strategy. I started listening to the crowd noise and becoming anxious to take off and go. Breaching my strategy, I began to pass Jack, but then pulled back remembering the plan.

Now as I faced the final lap, it was time to execute on

the final phase of my strategy. I took off sprinting as the bell rang alerting us of the last lap of the race. My father roared around the track along with my teammates, yelling "Go Rick, Go Rick"! The surge of enthusiasm in my heart began to build. Passing Jack, I never once in my final lap looked behind to see where he was. On this day, I had a plan, a strategy to execute, and the discipline to remain focused. I was the first to cross the finish line with a time of 9 minutes, 42 seconds. "I am the champ!" I remember yelling.

However, my team ended up placing 3rd overall. If I would have run both races and placed well in both, we would have certainly had a chance to win the entire championship as a team. There were no frowns by my coach or teammates, however it was the leadership value in me which echoed the truth that team accomplishments were far more important than individual accolades.

Nearly 30 years later, I'm still the number one runner on that day. But I missed creating a memory for an entire group of young men in which 30 years later we could all still be at the top of the heap. More importantly this was my Indian River High friends. Much love, guys!

This was my childhood breakthrough moment; it really characterized who I am as a leader today. I never again wanted to take home an individual prize without bringing home the ultimate team reward. I also realized having a plan for my life would become essential as a professional leading teams.

So, I'm excited to have you as my teammates, sharing my experiences regarding what I believe are essential principles in becoming emerging leaders for winning teams. I invite you to stand at the starting line with me as we work to complete a journey of leadership excellence and finishing together as a team!

INTRODUCTION

Leaders in today's fast-paced and ever-changing environment face chaotic experiences. The days can be filled with frustration and joy all in the same moment. Challenges and responsibilities placed on those in leadership are overwhelming, but necessary for the success of companies.

The expectation is to satisfy employees, customers, your boss, church leaders, head coaches, principals, city officials and frankly every business partner with an urgent need. Not to mention our significant others who are waiting to be fulfilled with the same level of expectations we commit to the workplace as is experienced at home. It's tough!

How can you effectively lead your teams?

How can you grow relationships at home when your days are filled with continuous change?

Compound the effects of children or community responsibilities, one could argue there is a full-time responsibility to please others nonstop. A towering climb of obstacles seems to always be standing in the way.

How do you invest in becoming a better you?

How do you lead with confidence?

How do you influence your teams and the people around you?

How do you maintain a healthy work and home life balance?

What are the most effective methods by which to lead in this demanding world of change and chaos?

The situations described were about me, Richard Mike. With nearly 20 years of experience in management, working within Fortune 500 companies and venturing out to start my own business, I've experienced first hand those towering responsibilities that become such a demand. I'm also a husband and a father. I've been active within the community as a basketball coach with my son for nearly ten years. I truly can relate with the demands of trying to please so many people.

Though I've always felt quite successful in leading teams, I recognized the challenges. The people demand has become a "deliver now" expectation. Whether it's the customers your business supports, the AAU parent who wants their child to start and score the most points, or even within the church, the competing desire to be the lead vocalist—everyone wants results now. Now let's add in the compounding advancement of artificial intelligence. It's obvious that the pace of change in our world of business has evolved with continued digital and technological improvements becoming an overwhelming focus.

Frustrated and feeling helpless at times, I fell into the trap of ineffectively leading my teams and more importantly my family. I struggled in managing my time, balancing both the demands of trying to be a successful working professional and trying to be the best dad and husband I could be. At one moment I may have delivered a kick butt presentation, winning over my boss. But I could only celebrate for a quick moment as my wife sent me a text reminding me to take my daughter to swim lessons. It's my turn. Damn, I had totally forgotten and was going to be late!

The balance was tough. The routine started all over the next day for my son's basketball practice. Managing life situations seemed to become harder both at work and at home. It just seemed there was not enough time in the day. I wanted to be super dad, super husband, super career guy all in one!

I discovered that many professionals within my community and colleagues in various companies were experiencing the same struggle. I was not alone. With companies all competing to meet the demands of their customers, they too were facing change and processing changes at an alarming rate. It took me months of working with companies to recognize one common ingredient impacting the ability to lead teams. CHAOS! Chaos was occurring by means of the very distractions we have created through cell phones, social media such as Facebook and Instagram, email, text messaging, youtube, and the expectation that saying "yes" would get in the way. Chaos has formed a roadblock in our ability to effectively lead.

As a result, I created **The Mike Method (TMM)**, a

soft skill culture of leadership that allows you to express with conviction a series of leadership principles beginning with the Letter "F."

At the moment of losing control of your day when frustration begins to boil, application of the **TMM** principles will help you gain control and unleash the talent in you and others. Instead of being drawn into the frustration bubble of chaos, you can divert your attention to leadership terms that begin with the letter "F". The method guides you in leading the people aspect of your day. **TMM** will empower you to retain winning, successful, and loyal teams!

In the May 2018 Harvard Business Review article, "Automation Will Make Lifelong Learning a Necessary Part of Work," it was stated that leadership demand will be raised by 24% in 2030. Soft skills and emotional connect ability will be required essentials to lead effectively. Your accounting, analyst, computer programming job today will be replaced by some form of Artificial Intelligence. Strong leadership will become valuable and essential within the workforce.

Leadership guru, Simon Sinek summarized leadership today especially for millennials as a requirement to invest in a technology focus, purpose, impact, and ensuring employees feel valued. Sounds straight forward and logical. Now you need a method to lean on to help discipline your day to be effective leaders. The **TMM** leadership method is your solution!

TMM was not created to help manage numbers, strategy, tactics, or technology. TMM was designed to help strengthen your people leadership skills. Through a series of principles that start with the Letter F, you will now take what today may be a level of FRUSTRATION regarding creating solutions and discover an easy method of achievement.

Many of us have likely been told, "Don't say the 'F' word!" It's not acceptable. Professional. Appropriate. Respectful. And I would most certainly agree. But once you read through The Mike Method "F" Principles, you will have my full endorsement to yell out the "F" word as loud as you wish. If you're ready, I'm going to introduce you to 12 F words to use! 😊

DON'T HATE THE F WORD PRE-THOUGHT

What Frustrate's you about leadership?

Before you go any further, Richard ask's that you take 15 minutes to think and write below those concerns that frustrate you about leadership. This goes from leading your family at home, church, office teams, sports program, school faculty, including any people driven environment. More importantly, what about leading you! Richard firmly believes your frustrations will be resolved as you navigate through this GPS guide. However, this step is critical to your journey. Be honest. Be expressive. Be prepared to lead winning teams!

1. _____
2. _____
3. _____
4. _____
5. _____
6. _____
7. _____

TMM PRINCIPLE #1:

FOUNDATION –
IS THE GROUND SOFT
ENOUGH TO LEAD CHANGE?

Preparing Your Foundation for Change – Personal Story

My family moved to our new home in the summer of 2016. This was a critical year for me and my family as we transitioned from Chesapeake, VA to Charlotte, NC in an exciting career change. Driving home was a joyful experience, mainly because our two children helped to select our home. I still wonder how we allowed that to happen. It's interesting how a four-year-old daughter and ten-year-old son could have that much of an impact on a home buying experience.

After a day in the office or a meeting with clients, for me it was impressive to see a neighborhood of homes with well-manicured yards. As neighbors we took exceptional pride in keeping a seasoned green color of thick fescue grass in our front yard. I took personal pride in my yard; I wanted to show that I could keep a nice-looking green yard like my father-in-law.

However, if you visited my home and made your way to the backyard, there was an opposite experience. Our yard was a composite of hard clay, weeds, and spotty areas of crab grass. We had rocks and leftover bottles from the builder that rose to the surface when it rained. My 4-year-old daughter at the time refused to go into the backyard to jump on her trampoline because the ground was so hard.

Being a sociable husband and father who enjoys his outdoors and entertaining guests by cooking out on the grill, my wife asked me one day, "How are you going to entertain guests in our backyard with an unattractive yard? It's a mess."

I didn't think a yard should prevent me from having guests over for a cookout; however, she was right. It was time to

My son playing football on a hardened foundation

swallow my pride and figure out how I could transform the look and feel of our backyard. My family, like our customers, did not want to enter into a bad place or experience.

Time for a Transformation

A month later, I looked into the possibility of purchasing seed and fertilizer to begin the transformation process. I even looked at purchasing an in-ground sprinkler system similar to the front yard. My thought was water, seed, and fertilizer could easily do the trick along with aeration Sounded good enough and made perfect sense to someone who was not a lawn expert. At this point, I was ready to start the infamous "project backyard makeover!"

Before beginning the process, I met with local lawn experts for estimates. I learned there was a key piece that needed to be resolved. An expert shared with me that the *"foundation"* of my lawn was not going to allow the growth of grass or the lawn restoration I was seeking. I would experience potential short-term benefits and likely remaining pockets of spotty thin grass throughout my lawn.

Without improving the foundation and establishing my lawn's core functioning capabilities, I would have wasted money.

My target state and vision made sense as my goal was to incorporate a transformational change for my lawn. However, like many of us today, we take a step forward in implementing a process change without understanding the foundational requirements to allow our teams to absorb, understand and respond with rallied support in executing the vision.

The solution the expert recommended was to break up the hard clay by tilling the yard to delve into the foundation of the yard. The tilling process required the over-turning of the hardened clay, removing rocks and waste matter. Then the top soil could be placed since the foundation had been softened. This would then create air flow to penetrate through the ground. I trusted the recommendation and decided to move forward. I'll bet you were not prepared to listen to a lawn care tutorial but hang in there with me. It's going to make sense.

I heard the late Dr. Myles Munroe state the fact that in order to improve the foundation you need to get under the earth. "Project backyard makeover" kicked off and the waiting period began.

Now let's put this in perspective for us as leaders managing a team, department, organization, you name it. Does this relate? Do you find yourselves often in this

Prepping the foundation for change

911 level of urgency, wanting immediate results? This was definitely me!

Tilling and new seed/fertilizer complete

Immediately, no patience. I called within 72 hours, questioning the process. I did not have any patience. I only saw myself staring at a lawn filled with what appeared to be mud covered with brown hay. I suggested to the gentlemen who facilitated my lawn makeover maybe we needed more seed and that the heavy rain washed it away. My wife said, "Honey, be patient!"

Here we were on day 5. I'm looking out my backyard window still wanting results. I was running water. My backyard looked a mess. And I wanted to see green grass. I called back again. "Hey, I understand this takes time. But I'm worried about the heavy rain that may have washed away the seed." The gentleman you could tell was getting a little frustrated. He said, "Mr. Mike, would you like a refund and for me to put the old grass back in your yard?" (in a joking way, of course). He reminded me to be patient.

On the 7th day, I began to see baby grass beginning to sprout. I remember calling the lawn expert saying 'guess what?'

He said, "You see grass, don't you."

I said yes! He chuckled and said, "Because the ground was tilled and the foundation of the lawn has been revived, simply be patient and allow the maturation of your lawn to grow." He said, "Continue to water your lawn daily. In combination with natural rain, this will complement the health and growth." I finally shut up and simply listened and trusted the plan. I realized that he sounded somewhat like me when my manager mentees are struggling in driving team results. If you recall my childhood breakthrough moment story, planning was not my strongest asset until desperate times called.

Within 30 days, I had nearly a covered backyard of dark green grass. Within two months I had a thick and fully covered yard of green grass. I remember my in-laws had visited over the growing period and were both amazed at the speedy maturation of my lawn. I was in business and ready to have my first cookout invitation. My mom then visited and was like "Wow! How did you import grass into your yard so fast." My mom was another one who joked about my backyard. Smile, mom! ☺

Reminder: we started this process around mid-September when the weather was in the low 80s and mainly 70s once October came around. What was important was that the cultivation of change was done at a time when the environment was ready to absorb the change. Stay with me as we tie this principle together in leadership.

I learned that the key foundational requirements of growing a healthy lawn was healthy soil, good air flow, water, nutrients, and sunlight. It's no different in leadership: your company, school, or sports team needs core fundamental beliefs to establish a trusted foundation for their teams. As leaders we need to create the right leadership nutrients developing a positive team environment that can help thrive through change.

So often in leadership we fail in the execution of

our performance. We are quick to roll out changes that often yield only surface level results. This means we are only getting partial effort from our employees as their foundation has been hardened. They don't feel a part of the team, but just a part of the process. As leaders, you are responsible for influence in a competitive world where the people you serve may not be as loyal as maybe 30 years ago. You find yourselves reacting quickly, wanting to make knee jerk decisions. Similar to the scenario with my lawn, I was ready to make more impatient changes. I could have easily damaged my lawn by applying too much change. Trust me, I would have doubled up on the fertilizer if I could. When reacting, we may easily find ourselves having to repeat work due to short-term fixes that leave your teams unresponsive and unmotivated. And you will likely continue to race to react, as urgency is a part of today's culture. However, in laying the foundation appropriately, you can feel confident that the seeds have been planted and you can trust the

My daughter and our dog now plays in our backyard effectively prepped for change.

process in allowing your teams to grow over a desired window of time.

Guess what. This works in all team people-oriented environments: Schools, sports teams, churches, law firms, medical teams, tire shops, restaurants, the list goes on. If you are in a people-driven environment, you will need to establish the right foundation for your team.

How do we ensure that the right belief value systems are in the forefront?

When working with leaders, a typical question asked of me is, "How can I help improve the performance within teams?" Another question I often receive is, "How can I help improve a broken process?" "How do I keep my strong talent from leaving?" I believe these are great questions which reveal common areas upon which many teams within companies focus. Performance and process improvement are certainly critical needs to remain relevant in today's competitive market. The strength of your teams tenure can do wonders in building trust and credibility.

Our consumers remain loyal based on the quality and speed that companies deliver on their products and services. We are in this world of delivering now and consumers like it. Ask former CEO Jeff Bezos of Amazon, as he has transformed the customer online shopping through a "service now" experience.

Before I even jump into wanting to fix the process or problem, I want to understand what is the culture of our team. What is the belief system of values that all members are to stand on? This is what's most important to me. Guys, think about when you meet a woman and you both fall in love. The girlfriend wants you to meet the parents. Typically the mom and dad want to know what??? What church do you go to? What's your faith? Where do you work? How long have you worked there? Have you ever been arrested? Okay, that's a bit extreme. But seriously, the parents want to understand your belief system before they fully endorse you taking their daughter as wife-to-be.

A common mistake I've made in working with a group was jumping right into the process fix. Let's admit, for many of us who think we are smart, we enjoy the challenge of fixing things. In my former business, I remember spending three days looking at data and developing process flows to understand the company's operating model. I was hired to improve some of their processes and I only had one week to complete my assessment and offer a recommendation. If you ever heard of analysis-paralysis, that was me. I was spinning my wheels re-engineering the numbers and processing so it would make sense. But something was missing!

I pulled their leadership team into a brain-storming discussion where we found performance gaps related to technology and training. Their workflow was extremely manual. The next move was to implement an automated pipeline system to allow calls and emails to be sent and work processed at a higher efficiency rate. There were improvements though we still were not seeing the results that I projected.

After hearing some teammates grumble on the floor about not being happy at work, I asked the organizational leader if I could spend some time with the teams. Some I met one on one and met with others through a roundtable open talk forum.

This is where I meet with the teams only and ask questions to create open dialogue and understand the team needs. I determined several teammates were not pleased with management. Uh, Oh. I should repeat this one. Teammates were not happy with their management. From issues of pay, failed communication, constant change with no reason, teammates were not motivated to work. So even with the recent automation, they did not build a good "Why" statement regarding the change which was being implemented. Based on the feedback, this was the culture of the company.

Let's take it a step further, even within the sports arena. We as sports fans are only caught up on the output of watching sports. Rightfully so, we are not at our favorite team's practice, we've got our own lives going on. We only care about watching our team (like me, the Dallas Cowboys) on Sunday. Go Cowboys. The same situation occurs. New coaches come in. Some dive immediately into changing the entire scheme and culture of the team. No deep dive into the issues or understanding of the player concerns. It's a macho play. The coach has his or her own thought process of what works and jumps in to immediately drive change. Oh, and what about those situations in which the general manager doesn't even talk to the star player about perhaps a key player move. Is it necessary? No. But you will quickly lose the trust of the team if they don't feel engaged.

Going back to the issue at the office I was consulting, I

pulled the leaders into a meeting and stated that they needed to address the people aspect of their company. "You have no core values. Your managers have no guide to lead and no foundational culture to influence their people. Until this is corrected, you can pay me to continue implement new processes but you will never get the full output. Right now your morale level is below zero."

This was a lesson I learned even in my corporate career. I often wanted to jump in and look for the sexiest process change to gain kudos from my boss. However, if the people are disgruntled due to issues that have not been addressed; and as leaders, if you are not living the core values of the company, how good really is the process fix? Will your foundation truly be ready to inherit the growth without the core value nutrients to keep your teams grounded?

Foundational TMM Beliefs that work for me

Unfortunately, what I'm going to share with you will not be the latest breakthrough leadership philosophy of setting values. You may have seen listed values within your own company, school, basketball gyms, church, local doctors' offices. The list may even hang up on the walls, very visible in an effort to remind members. And if this is news to you and sounds like ground-breaking values, even better. Let's do it!

Communicate with transparency

I get it. We are in a different time today where effective communication becomes one of those questionable areas of what works. With technology, I believe communicating with our teams at times has become a lost art. Too often, we speak through typed words with emotions that get missed – wrongly interpreted. We sometimes rely on electronic tools to avoid the personal nature of sharing difficult decisions. We are afraid of tough eye to eye discussions.

Good communication creates transparency for you and your teams. The more you communicate, and the more you create an environment that fosters two way dialogue, the better your team will be.

Nothing drives me crazier than an evaluation you receive in which 75% of your evaluation is new news. You are sitting there like, "Where did all this come from? Where is the transparency? Do you think maybe you could have given me at least a small clue of how I'm doing? As a Jr. Pastor, you are thinking you're next in line to pastor your church. As a college sophomore shooting guard, you're thinking you are going to start this season. As the lead nurse, you are thinking you'll be promoted to nurse manager. We'll spend more on this later. But you get my point. Team morale will rely heavily on your communication and transparency. I personally don't think this is rocket science but a common piece that's missed.

Promote Change As a Positive

Growth does not happen without change. GROWTH DOES NOT HAPPEN WITHOUT CHANGE. If I was writing a book solely on understanding change, I believe I could get away with repeating this statement over 1000 times and I may end up with a best seller! GROWTH DOES NOT HAPPEN WITHOUT CHANGE. Creating a mindset and culture around change and its value to the teams and customers builds a winning experience. Resistance to change impedes growth and performance. It doesn't get more straightforward than this.

The ability to lead change begins with you as the leader. We'll spend time on this aspect later in the book. But just know, your teams will only be as flexible as you are. If you are stubborn regarding change, and want things to be steady without even a wrinkle in the road, you can still get the job done but you are likely not going to be a high performance leader with impact. If you are in a current leadership role, you are likely failing or mediocre at best. If you don't like change and want to go with the flow, your role will not last long. A new coach, manager, principal, will soon be taking your seat.

I define change quite simply: a move from complacency. Think of a home designer. How successful would home designers be if they only created the same single home interior layout for every home in the neighborhood? Could they get the job done? Yes. But some consumers are likely going to want something different—something that stands out against the norm. So to continue to remain vital and reputable in the business, the home designer must study their market demographics and truly understand what it is that the consumers are looking for. Are the 12 foot standard ceilings becoming too much of a norm when consumers may want the high 30 foot ceilings, creating an open space of luxury. The point here is that as leaders we must always be anticipating change based on the demands of the people we serve.

A great example could very well be the COVID 19 pandemic that has impacted our world. Surely you've noticed the profound impact this is having upon our youth and the educational adjustments that are or are not being made by the school system. Realizing the market impact, schools are investing in remote learning, understanding that this may be a realistic form of learning for years to come.

Change creates an expanded opportunity to learn and achieve greater heights of development and success.

At the end of the day, if you want to lead people become a fan for change. Remember, you as a person went through some type of change from birth til now. You grew. Don't you want your teams to grow too?

Build Authentic Relationships

It's taken me nearly 20 years to figure this out: Create a circle of people around who can **HELP** you. The key word here is **HELP**, bolded and in caps for a purpose. How many of you still today are trying to accomplish too many things on your own? From parenting, home and car maintenance, cooking, exercising, starting a business, investing, financial planning—these are just a few areas where often we try to do too much on our own and will not ask for help. And to be honest, we suck at some of this stuff! I can't fix a damn thing in my home yet I'm always trying. LOL.

For those that do ask for help, job well done! You get a gold sticker. There are still individuals like myself who can be a bit controlling, carry a big ego, and still at times live in this perceived world of "hard work makes the dream work." Well, I challenge that cliché. What I hear nowadays is, "It's not so much the numbers you make, but the hands you shake!" Sounds corny, but what the heck was I thinking all these years refusing to ask for help?

I started a coaching business a few years ago that failed quickly. I was excited about my ability to train and coach leaders and felt I was ready. My biggest failure. I did not have the right relationships. I was doing everything on my own. Networking events, creating my own training and consulting material, social media posts, you name it. I later found I was not good at networking. I should have asked for help. I was like a misdirected bull in my introduction, jumping straight to the sell immediately. When I realized I had completely sidestepped relationship-building, I could only shake my head.

As leaders, I strongly encourage you to focus on building relationships with your team, even with those surrounding your team. The reality here is that you will need help regarding how you lead your team. You will need an outside support from a peer group within your company, organization, etc. If you have the right relationships, your help will always be on its way!

Trust and Confidence – Everyone wants it!

Let's start here for those lovebirds who are married or thinking about marriage. Here is a popular one-liner from one of the many wedding vows out there. "I promise to love, respect, protect and TRUST you." When you think of a union of two people, there is this element of trust that plays a key factor in the strength of the marriage and its ongoing growth. If the trust is broken—and you hear this all the time in struggling marriages—the relationship is just not the same. It may take several years to regain and/or there will be ongoing animosity that eventually destroys the relationship.

Guess what? Your teams want your trust. They want to know that you, as the leader driving their team forward, can be trusted. If you have a team that is going to work long hours, endure change, deliver top tiered results, do they have the trust and confidence in the person that is leading them? This is critical. If trust is missing, your teams are not going to bat for you. I'm not saying they will walk out the door—again, like a marriage, but they may do just enough to get the job done but you won't get their full buy-in. You won't hear their ideas or see their talent on display because they are no longer on board with your plan. The trust is broken.

Something to keep in mind here also. The generation of people have changed from the student, the parent, and the employee. Those from the baby boom generation grew with a hard-nosed work attitude and understood that they must endure giving all they have to their job despite whether the boss is a winner or a loser. These days, if you are not trusted, folks will leave you. I see it in high school sports. An unreliable coach or teacher, the parent leaves the school with the son or daughter. An untrustworthy boss, teammates will leave. It's a different day so unless you want to deal with continued turnover and loss of talent, be a leader that is trusted.

Love Your People

In sports, you often hear this cliché used by coaches and team owners, "Trust the Process." It sounds catchy. I beg the differ. I say, "Trust the people within the Process." It is important in leadership to not fall in love with the process, strategy and administrative stuff, but have a genuine care for people. Your teams play the most integral role in carrying out the vision. In our ever-evolving global economy, process innovation is prevalent. As we glimpse things like Google's Alexa and Tesla's autonomous cars, the technology is captivating. CNN, Wall Street, and USA Today will certainly find ways to report on this type of technology advancement. It's eye catching with a "Wow" factor and customers thrive on the excitement.

But don't let the people aspect become a low priority. Unfortunately, I've seen many managers, including myself, become so absorbed with projects and problem fixes to the point where their teams become de-valued. Interesting, some companies do not de-value the customer experience, yet the employees at times can get forgotten. Don't allow that to happen; prioritize the people needs.

I have an uncle who constantly tells his family members and/or close friends that he loves them. As soon as he sees you and gives that big uncle hug, he immediately says, "I love you." I also play basketball on Sundays at a local church. The organizer of the church gym immediately greets you with a hug and a big "I love you!" I enjoy being around both of these men. You feel their genuine care and love for you so it makes us want to be around such individuals.

I'm not suggesting you run to every teammate or student and say, "I love you." But I have seen the performance of teams naturally increase just by those in leadership roles showing personal care and support for their teams.

Be a Game Changer

Look, your people will have the ultimate respect for you when you deliver BIG. It's that ah-ha moment that captivates your teams when you lead from the front and can drive exciting changes. Be creative and encourage creativity—to rally your people to come up with innovative ways of doing things better. Look at us as consumers; aren't we mesmerized by game-changing leaders? Jeff Bezos' Amazon online shopping experience was a HUGE hit. Steph Curry's (Golden State Warriors) ability to shoot 3 pointers from all over the court is game-changing. In view of the musical excellence and entrepreneurship of Jay Z and Beyonce, they have built their reputation as one of the more successful couples in entertainment, with fans all across the globe. Jay Z even said once that the 30s was the new 20s. He was right. They are game-changers.

I just believe when you deliver BIG, especially in important moments, you gain such an admiration and influence, people will want to follow you. Your game-changing thinking then becomes contagious!

So you have it. These are six TMM beliefs that work for

me. I've seen it. I lived it. I have experienced it in many aspects through both success and failures. Take and run with it, while using this as a foundational start in leading your teams.

TMM – Time to Ignite the Foundation within your team

- Develop or learn the existing core beliefs that represent the culture of your team or company. Remember you may have to dig beneath the surface to get to the root of what keeps your foundation stable.
- As a leader, be the example for others to trust and follow. We've all heard it. Walk it before you talk it. Actions speak louder than words. And my favorite when I was a kid: "You do it first."
- TMM: Beliefs that you can leverage if not already being implemented today:
  - Communicate with transparency
  - Promote change as a positive
  - Build authentic relationships
  - Demonstrate trust and confidence
  - Love people
  - Be a game-changer
- Implement your core principles within your project, meetings, playbook, etc. Keep it visible so everyone is reminded, especially you.

I promise you, if you do the above within your existing values or implement belief values that work for the culture of your team, you will be miles ahead in leading your team and being the best leader you can be. Because when you hit that wall of frustration and are not sure where to turn, you have the F word: "Foundation" established for you to settle back into to keep your team moving forward.

Y-G-T You Got This!

TMM PRINCIPLE # 2:

FIND IT! FIND THE TALENT IN YOU. FIND THE TALENT IN YOUR TEAM.

It was Friday, November 25, 2017 where I held a TMM Ignite forum. This was a networking opportunity with a small group of successful professionals and friends in my hometown of Norfolk, VA. Our topic was on personal growth and the power of networking. We kicked it for about a good hour while eating appetizers and, of course, imbibing a couple rounds of drinks. It was good for us to simply catch up and talk. We finally kicked things off on a personal note as we each spoke and gave a 10-minute overview of our careers, personal journey, fun facts, and shared our passions of interest. The discussion far exceeded my expectations. I was impressed with some of the backgrounds and stories that these great men revealed.

What was most exciting? No one was complacent with

their current success. Each person shared ideas and new opportunities that would drive their growth mindset going into 2018 and beyond. Our conversation grew deep. We'd crossed over to a place in which we were genuinely open about life. It was on point!

Dr. Ricks, who teaches leadership within his professional career, was our keynote speaker at the event. Dr. Ricks and I worked together early in our professional careers about 20 years ago. Our friendship quickly connected. We both played basketball, were goal-driven, and were just two silly guys who could always engage in good conversation and laughter.

After 15 years, we connected again over the phone in June of 2017. We still shared the same interests—silly, goal-driven but now we could talk about being husbands and fathers. Wow, has time flown by!

Dr. Ricks left us a message pushing mind thinking strategies, influencing change, and the importance of knowing our strengths. He passed out the book titled, *Strength Finders 2.0* by Tom Rath. This book was specific to understanding our strengths and talents and how they should be utilized within our journey of growth. He really sparked an inspired sense of understanding how important it is for us all to live to our full potential. Each person walked away that evening so pumped up, ready to take on the world with enthusiasm, knowing we had genuine support from one another.

The next morning, I was driving back to Charlotte, NC and I thought, "Damn, that was a powerful event!" This was a six-hour drive, so I had plenty of time to think and reflect. I was so touched by the event, I literally replayed every aspect of it from the venue set-up, conversations, the 10-minute presentations, and Dr. Rick's message. This was an outstanding conversational forum, not to mention the drinks and wings were "banging," as my kids would say!

I kept replaying the talent portion and I was anxious to read the book. So that Saturday evening, November 26th, when I returned from Virginia, I began with great excitement reading my *Strength Finder* book. I was anxious to FIND the talent in me. The author invited readers through a reality check in recognizing strengths, especially as it related to goal setting based on our passion. After reading the book, I took the assessment, a 20-minute survey of 150 straightforward questions. Based on the answers, readers were given a summary of our top five strengths. For me, the characteristics align with who I am today as a leader. My key strengths came back as follows:

> Individualization – A keen observer of other people's strengths and talents, drawing out the best in a person.
>
> Communication – Very good at piecing stories together, presenting to others and being a good conversationalist.
>
> Relator - Very effective in building impactful relationships, understanding people's feelings, goals, dreams, fears, while sharing my own.
>
> Arranger – A conductor with a natural ability to

orchestrate people and resources for maximum effectiveness.

Belief – A believer in core values is critical. It draws me close to family, spiritual themes, high responsibility, and high ethics.

Though these attributes fall in my areas of strength, seeing it on paper was like staring into a mirror. This was me—it was who I am. These are the God-given talents I've been blessed with, so I need to flourish in these areas at all times. Maybe if I showed this to my wife, she would understand why I talk so much and why I enjoy hosting social gatherings at my house. Smile, Mrs. Mike! 😊

I thought, if my talent is to speak and help others, I want to become the best damn speaker and coach I can be!

I started thinking about my professional career. Why didn't my managers know that these were my key strengths? We invested a lot of time both on coaching and process improvement. But I don't believe people really knew what my key talents were. Yet, maybe I didn't even realize my key talents and/or how to utilize them in areas to help my organization. As a result, I wanted to always put myself in a position where I could motivate and coach others. Again, I wrote this book in order to motivate others with principles that have helped me become a better leader.

Then I thought about teams I've led in the past and teams I've watched be led. I shook my head as I realized things I had done wrong. If one person stepped up big, I ended up trying to use this person for everything—communications, research, coaching, whatever. I would work the heck out of the person. I didn't realize that I'd roped individuals into presenting coaching sessions when they weren't even gifted communicators.

I think about sports teams. You watch some games and you just wonder why certain players are playing positions that just don't make sense. The player is fast, quick, shifty, has great vision when he runs so you think he should play running back. But the coach lines him up as a wide out running deep routes only. Head scratcher! You have a basketball player that always wants to score when he has the ball, yet you have him playing point guard, which is a floor general role.

At church, you are using Ms Mary to do your accounting and she still wants to print brochures only. Yet you have a young lady who is a social media mastermind wanting to volunteer but you only use her as a door greeter. You have to know your talent and know the talent within each member of your team.

Imagine the dating experience. These days of social media and video phones, much of the initial dating may be done over the cell phone or through skype. I'm not sure I could survive dating in today's time. But think of how many dates you've been on in which early on the person states they want to know about your talents and your passion in life? They want to hear about your dreams so they can support and help you in the process. You probably rarely get this level of attention when dating. Especially we men are probably not asking those questions.

But wouldn't this stand out and at minimum make you

stop and think, wow this person may actually care. It's the same concept in leading teams. I've helped teams by using this type of talent tool and it has helped managers gain more trust and confidence from their teams. "Finding" the talent within your teams means you care about their success and how they can contribute to your ongoing winning success. First, it begins with finding the talent in you. But guess what? You may go through your talent assessment and realize leading people is not your gift or passion. And it's quite okay. Listen, we all have taken on responsibilities that may not comprise our greatest skill set. But it helps you become well-rounded and develop skill sets and know-hows in other areas. So, once you do get into your role of passion and self talent, you can bring these added qualities which can only make you a superstar.

More importantly as leaders here is where this gives us an advantage. Based on your talent assessment, you must know your strengths and the strengths of your teams. This allows you to now DELEGATE. One more time. DELEGATE. Leaders don't do it all. Leaders understand to maximize the potential within their team. This only builds that Core Belief of "Trust and Confidence." The more you delegate, the more you are building a psychological brain stimulator within the team. They feel engaged and of great value. They are excited that they can use their talent to help you and the team.

If you are FRUSTRATED with so much going on and not enough time in your day. Use the TALENT within your team. Coaches. You have assistant coaches. Allow them to dissect the film review in detail. Pastors, allow your Jr. Pastor to talk about faith in today's time. They may be able to relate to your people more based on generational similarities and differences. The point: FIND the talent in you. FIND the talent in others. This will only build the overall within your Team.

TMM - Time to Ignite "Find Your Talent"

- Managers, you should check with your human resources area or training team to determine if you have a strength assessment tool
- Once you have the assessment tool in place. Use it. Use it. Use it.
- Complete one for yourself to ensure you understand your strengths. I'd recommend focusing on the top three strengths in the beginning.
- Have your team do the same. Meet with your team and discuss how their talents can be used. Honestly, I'm perfectly okay if the conversation shifts in the direction of whether if they are in the right role. Be a good leader and help find the right one.

TMM PRINCIPLE #3:

FUNDAMENTALS – SOME THINGS YOU MUST GET RIGHT EVERY TIME

Growing up as a youth and throughout college, I was involved in sports. Since the age of six, I played little league t-ball and continued with sports through high school and college, including basketball, track, and some football, like most kids I grew up with. Well, football I scratched in the 7th grade after I got gang tackled on the sideline and I saw butterflies hovering around me. Justin and Terrance, you got me good! That was it for me after that season. Ha-Ha!

I spent my childhood growing up with several of my friends from my neighborhood and school who were all competitive athletes. Several of my friends earned college scholarships in their respective sports, largely driven by grit and a determined mindset to compete at the highest level. This included me. Quick shout out to my Indian River family! I ran the 1600M and 3200M in high school as I noted in my introduction. I ran the 1600M and 3200M in high school as I noted in my introduction. I eventually competed in track and field at Norfolk State University (One of the larger HBCU *Historically Black College Universities*) in the country. In 1994 we won our first Division Cross Country Championship in school history! I had to call that out. Go Spartans! But overall, sports has been an amazing experience for my growth as a competitor and, more importantly my development as a leader.

Learning the importance of fundamentals early

In my 8th grade year 1988/1989 at Indian River Jr high school, we had nearly 20 of us on our basketball team. This was not your normal roster, as most teams kept 12 to 15 kids. But we were a talented bunch led by one of my good friends still today, Justin (we called him the young Michael Jordan). Yep, the same kid that gang tackled me. LOL. He had the athleticism and finesse to finish at the basket even in the most awkward positions, hanging in the air with amazing flair like Jordan. I had such an amazing appreciation for Justin's game as he seemed so advanced beyond our years. It was really incredible watching him play. However Justin's emphasis was the same as the remaining team members in that he placed a great deal of effort on the fundamentals. Justin would go on to play NCAA College football as a Safety and today is a successful entrepreneur and co-owner of Great White Water Sports. Way to go Justin!

Recognizing the explosive potential of our team, our

coach would really invest his time in preaching fundamentals. I recall there were two days of practice where we wouldn't even pick up a basketball. What 13 and 14-year-old kid wants to come to practice and not have a basketball in his hand.

We would run endlessly, constant footwork drills, defensive mechanics and shooting techniques without the ball. We were like, damn, is this track practice or basketball? He finally introduced the basketball on our 3rd practice. It was the simple fundamentals of dribbling, passing, defense, and layup up drills designed to enhance the basics of basketball that could be utilized in any offensive or defensive scenario. To earn playing time in the game, you could not miss layups, turn the ball over, or lack defensive footwork. Shooting a long 3-pointer was not a fundamental area of emphasis. In fact, a player may get pulled out for jacking up dud 3-point shots. I could shoot that 3-pointer but I would get pulled to the bench for not playing good defense.

We finished that season with a 13-0 record, winning our 1989 middle school city championship. Without a doubt we had significant talent. But fundamentals were what made us great. Great job, Coach!. Thanks for the fundamentals, even though we had some boring practices. We needed it!

It makes sense for any successful team to thrive, there's got to be basic fundamentals that you as a leader have to get right every time. Your foundation is established. You know your talent and the opposing teams. Now it's about delivering on five basic fundamentals I believe are "can't miss" routines that are essential to your success.

Let's Go!

FUNDAMENTALS – SOME THINGS YOU MUST GET RIGHT EVERY TIME

Practicing Good Team Fundamentals

*These are "must do" actions from your
teams to help drive performance*

Fundamentals are typically not the flashy objects that get our attention. Fundamentals may not even be the fun aspect of our jobs. However, when you master the fundamentals, the shining results will follow. The following represent basic leadership fundamentals I believe we must get right as leaders.

1. **Know your Playbook –
   Do it the Xbox / Playstation Way**
   Having children opens the door to several examples that I can share in terms of having a system everyone needs to know. Especially as they get older over the years, I see so many work- related scenarios. When I used to coach my son's 12-year-old AAU basketball team, my son Jordan and our entire AAU basketball team had the PS4 or XBOX system. They would play a popular game at the time called Fortnite. Fortnite had become the thorn in my side as my team would play this game all night if you allowed them to.
   Here's what I learned about the intelligence of each of my players. This is a game very similar to mortal combat back in my day where you go through some adventurous territory and basically look to shoot down the enemy. The game is meant to be played through a headset competing with other teammates from other households. To be good, you have to know the system. You either have to play

multiple times all day to become efficient or go through their training demo to learn moves. My son even showed me at the time a youtube tutorial they would watch to learn techniques and strategy.

I later thought to myself: dang, this was life. Real world work-related stuff. In leading your teams, you must design a system that everyone has to know. They have to be damn good at it. If you work in a surveillance monitoring business, make sure everyone understands how and when to use some of the computer devices to detect and respond when an emergency strikes within a home. No short cuts. The worst scenario can occur. You need something done and someone doesn't know how to use their system or know where to go to use it.

Think of NFL football. I talked to some friends I know who played in the NFL and they shared the fact that it was a fundamental expectation to know your playbook. The playbook both on the offensive and defensive side of the ball represented instructions to follow for all sorts of game time scenarios. Some said they would have to memorize a 100+ page playbook. Sheesh. But their explanation was that their coaches felt that if each individual knew the playbook, the entire team could be in sync when it was time to make real time adjustments throughout the game.

The same goes for your workplace. If you are in a working environment where you need your teams to follow various scenarios, you should create a routine in which your teams are having to answer questions related to current procedures. Think about it. You are running a restaurant and you decide to change the menu for the day. It's Wednesday and today's lobster tail is going for 41.99. You take the orders and you're looking at the menu where the lobster tail is 54.99. You bill the customer 54.99 and the customer complains. You show the customer the menu showing the price. The customer shows you the deal of the day flyer with the special—poor experience and representation of the company in not knowing the changes. If the waiter had checked the order against their playbook (meaning price specials for the day) this could have eliminated a bad experience for the customer. Fairly straight forward example. Your responsibility as the leader is to ensure your teams have an effective playbook to follow.

2. Set the Example and show up Early

Sounds silly. We are basically talking about showing up for the job. You may think this is a given. And you are right. It should be a given but it's something I see quite often. I see those in leadership positions show up late or just barely on time. Listen. You are the leader so your presence and timing is being watched by your teams. If you show up late or just on time. Your team may show up even later. And for you young athletes out there with aspirations of playing professional sports. This is critical. If you watch the NBA draft, you will hear analysts talk about kids leadership who are gym rats. They are the first one in the gym and the last to leave.

When you show up early you immediately set the tone for the day. When you arrive early, you are likely developing a plan that's going to be successful for the people that work for you. It's a mindset thing. This goes for coaches too. I'm telling you. My son played for an AAU team, and on the 8:00 am Saturday morning games, the coach would show up right at 7:55am like clockwork. Heck, sometimes I would be out warming the boys up. Sheesh. And sure enough our team would get rocked!

Don't even mention going to the barbershop at times. I've learned to never get there at opening or set an appt for the first appt time. Why? The barber who is the owner is likely going to show up late. Just keeping it real. Me as a customer in this day of time have too many things on my plate and need to see promptness. By the way, my barber may come in 20/30 mins late and then the rest of the crew come sliding in. Too funny. But shout out to the barbers. We love you but I'm glad I now have a bald head so I don't worry about this too much anymore.

3. Save the email and walk by and say "Hello"

Talk about simplifying the lost art of communication and inspiring your teams. I'm amazed still today by those in leadership positions who do not start their day off in speaking to their teams—and/or at least making a point within the day to say hello. Heck, I'm concerned now with today's advances in smart phone capabilities and applications; the morning greeting may actually be a bot that says hello and gives the team a big greeting. I guess the newer generations may find it cool! But for now, we need personal connections until I'm proven otherwise.

Now, understanding what level of leadership you may be in and the number of team members, this may not be easy to apply. Now let's keep it to a smaller environment where the visibility is quite obtainable. One of the biggest issues I heard when I spoke to teams when I was consulting, was: "I never see my manager." Or, "He or she doesn't even speak to me unless a problem arises." What??? This is a huge issue for me and is unacceptable.

During a roundtable forum I conducted with a company, this was the first issue that hit the discussion fan. In my mind I shook my head. But then I thought, if this is the biggest problem they have, I have an easy fix to put in place. Get your butt out of the office and say hello to your team. Done deal. Next issue please. ☺

Lesson Learned!

Back in 2006 I took on a new leadership position with a company in which I had a staff of over 60 employees. The challenging piece for me was the fact that folks sat all over the place. And this was a huge office floor separated by walls almost like a maze. My operation included a multitude of functions so they could not sit together. I had teammates who handled mail, indexing, printing, letter preparation, took phone calls, and this was from 8:00 am to 11:00 pm at night. I soon recognized after my 3rd day on the job, I hadn't spoken to everyone. I would get so sidetracked learning the job and with the different shifts; I did not have a plan to keep up and stay visible. I remember one lady stopped me and said, "Hello, nice that you decided to come by today." Not a good way to start your first week. I had that stunned look of, "Yikes, she's right!"

I had to make an adjustment. Around 8:00 am I would

begin my office "walk the floor tour," stopping by each person's desk and saying hello and perhaps engaging in some small informal dialogue as we jumpstarted the day. Those that came in at 3:00 pm, I was back walking around getting my "hello" on. Just through this routine alone I quickly gained the followership of my team. I hadn't even deployed any new process changes, and we saw our productivity increase by 15% due to the high morale. They were just happy to have a manager that took the time to say hello. They were motivated!

4. **Communication Routine - Your investment in communication will increase your stock**

Don't we all like to be informed? We want to know what's going on. We want to know "why" things are happening the way they are. Being informed helps us in general feel like we are learning or simply raising awareness. Today we all have smart phones through which we are continuously fed with information. And we like it!

Brief childhood Story

As a 6-year-old, in 1981 my family had one television in the house. Our television was the big brown wood-finish siding that felt like it weighed a ton. It took two or three of us to lift that sucker. Fortunately, the younger generations have no clue what this television looks like. What I didn't like was being a small skinny kid trying to help my dad carry this monstrous thing up the steps any time we moved.

Our routine was consistent. My father would pick me

up from after school care, and we would then go pick up my mom. My mother was a banker at Bank of Virginia in downtown Norfolk, VA. This was back in 1981. Wow, that was a long time ago. We then drove home and began to settle down as a family. This meant my mom would begin cooking. I had to do my homework. My dad would watch both the local news station and National CBS news led by the late Dan Rather. This is where it gets painful. You only have one television and it's the month of November and it gets dark at 5:30pm. So, instead of going outside to play, I had to endure the pain of sitting on the couch alongside my father and watch the news. Local Hampton Roads, VA news was at 5:30pm followed by Dan Rather at 6. Then on Sundays, 60 minutes after the football game. TORTURE!!

It sure would have been nice to have smartphones and tablets back then to watch shows on Netflix. For my younger generation, you just don't know how good you've got it today. We'll talk about that later. Hang in there with me.

There were certain things I did not realize until later in life, as I became an adult and a father myself. My father, Evans Mike, was extremely intelligent for a man who did not go to college but went straight into the military after high school. It seemed every question I had, my father always knew the answer. He would talk about the Reagan administration at the dinner table as if he was literally in the White House. I thought my dad was a walking genius!

But what I later realized in life was that my dad was just very much informed on local and world affairs because of the news. He was able to truly engage in dialogue on so many topics such as local traffic jams, the weather, politics, sports, you name it. The news was his repository source of information.

More importantly, the news represented a consistent daily routine of communicating with its viewers. Anyone watching the news became very informed of what was going on in the world. The daily news represented what I consider an excellent forum for receiving communication updates. Just as a child, it haunted me in my dreams! Now today, you have Twitter, Facebook, LinkedIn, blogs, and smart phone applications that give us an added source of information to remain informed without feeling we are being locked in to the TV alone.

My point here is the communication routine. If you have a communication routine established—whether it's daily, weekly, two days a week, etc. this will help to ensure your team is engaged in the happenings of your business or specific field of work.

Here are a few basic and simple routines that in my opinion works for any group settings.

- Conduct huddles, have a couple of key agenda topics that are critical to the team. Performance details are always great. Perhaps discuss the market competition and innovative changes happening that may impact your team.
- Use communication apps. My son's coaches do a great job using a parent appl to communicate key information such as out of town hotel details, game times, volunteer needs, and motivational quotes. It

works great and it makes us as a parent feel very well informed and engaged.
- Have someone take notes and send them out to the team if applicable. Someone is not going to be there and maybe it's not every conversation detail but pertinent updates. Plus, you know, people seem to forget things quite easily regardless of how clearly your message was delivered.
- Switch up the format – You don't have to lead every communication huddle. Allow someone from your team to step up and take the lead. Your team is tired of hearing the same voice. I remember as a track athlete, me and some of the guys would start laughing in the background because we knew the same old speech was going to be repeated by our coach. My bad, coach!

If you have a communication routine established, like my father, your teams will be well informed just based on your shared information. Well done dad, "Are you sure you knew the answer to all of my questions as a kid?" Smile Dad!

5. Connect 1 on 1 for 2 way feedback

Think back throughout your career or even now perhaps. You were pulled for a performance review with your manager and it didn't go so well. You received that rating with which you did not agree. You are sitting across from your manager and he or she is telling you all those things you should have done. You didn't do right. They didn't see your initiative. You could have done more. And you have that surprised look of being caught off guard. You're thinking, "It would have been nice if they had told me this before my annual review." This was your pay raise that you got all excited about that just got slammed. Chaos!

I chuckle when I use this analogy. However, it seemed to always hit home with a smile as I mentored other leaders. If a manager was frustrated that their team was upset about their feedback. I would always ask, have you been consistent in providing candid feedback throughout the year. Often the answer was "No." This piece is critical especially if you are building an environment of trust.

This goes down in sports and in all sectors of business. Heck, even especially in youth sports, I've seen the frustration in kids from middle school to high school. Every kid obviously wants to start and be a key contributor to their team. The toughest piece is not every kid can be a starter or be a rotation player. The challenge for coaches is having that candid conversation as to why. This is the 1 on 1 feedback that is necessary. Not every kid is going to agree but the sit-down conversation at minimum shows you care and you are providing honest feedback. For the parent who gets upset because their kid is not starting, they can't say but so much if the coach is being honest with the kids in evaluating their performance. Tough position for coaches because you can't please everyone especially the parents! That's the tough grit of sports. This happens often. The point still remains. When you are in a leadership role, you want to win the trust and confidence of your team. These 1-on-1 discussions are critical.

If there is one piece about leadership that I'm the most passionate about it is the 1-on-1 coach sessions. This goes for all managers and leaders of people. You have to meet with your people. They need to understand how they are performing to ensure they're meeting your expectations. More importantly, your people need that opportunity to maybe share with you some things on their mind. They may tell you how much they don't like working for you so be prepared. Just kiddin! But the opportunity of meeting one-on-one presents the candid conversations that your team is often hesitant to share.

Dating Example – "You could have told me something!"

It's kind of like when you were dating and you found that dream significant other. Everything seemed perfect, the smiles and dinners, romantic walks on the beach and restaurant eating in front of a fireplace or rooftop restaurant. Then all of a sudden, your significant other gave you the BIG rejection statement. "I'm not interested anymore and would like to break up!" You were caught off guard, thinking he or she was the one! And the feedback given was that you snore loudly at night preventing your significant other from sleeping. It was a turn-off and it had been a turn off for months; however, you did not know. This apparently was a huge issue because there was no turning back at this point. And you were thinking, wow if they had told me, I could have done something to fix the snoring.

That may be a bit unrealistic; however in life we get

these impromptu rejections that leave our mouths hanging open with frustration. We lose trust and confidence in the people we believed in—partly because of the lack of 1-on-1 feedback sessions.

A few tips on effective 1 on 1's that work for me:

- Start with some type of positive intent. Whether it's regarding their specific job or their personal life; this helps to relax any nervous thoughts
- Ask open ended questions to get some conversation flowing. I'll be honest, I've tried this with my son and he may be one exception to the rule. But for the most part it works
- Find out what's important to them and not you. What makes them tick and what are their goals
- Discuss what the concern/issues are and why you as the leader want to help. Use facts. Remember that youth sports piece I mentioned earlier. I was that fired up youth coach dad that later realized I had to use scorecards to rate my kids basketball performance. My kids were smart. The data motivated them to get better versus me yelling all the time. Shame on me! LOL
- Generate ways collaboratively to improve or build long term plans to help reach a specific goal.

I'll stop here as coaching is truly an art and may require training. But this gives you a baseline guide of how to approach important conversations with your teams. I promise, you will at minimum win their respect if you have these types of conversations. You can do it!

6. Recognize as much as you can!

I love my two kids and if you have children, I'm sure you love your kids as much do my wife and I. Being a parent is certainly a wonderful gift from God and watching the personalities develop quickly from birth even within the first two years is amazing. What stood out to me was the influence of a high-five. When Jordan or Taylor did something good, we would give them both a high-five and they would just start cheezing.

It was like the high five was the greatest reward for accomplishment. When he learned a new word or ate the green beans that he did not like, he would earn a high five with a big smile! Unfortunately, the high five recognition technique faded away over the years. Since about age 10, Jordan would be looking for the latest pair of Lebron James or Kyrie Irving tennis shoes. And my daughter, a trip to Claires for her 10th cell phone case and Dunkin Donuts. SMH. I'll enjoy this for now until she becomes a teenager.

Where are we going as leaders? Even as grown adults in the working world, we appreciate the value of being recognized for the work we perform. What about those who say, "I don't need a lot of recognition, I just show up and do my job!" They are the main ones who appreciate the recognition— either that or they're not working well enough to be recognized. Let's keep it real!

Give some love to your people when they do great things. It's rewarding and it again goes back to that trust and confidence belief. You recognize you've earned their trust and they will continue to go to the limit for you!

TMM Time to Ignite your fundamentals

- If you are not a people leader then you are in the wrong business. If going out to speak to your teams is a struggle then you need to reset what's important. Save that email for later and prioritize your people. If you feel time is limited, now it's about incorporating a routine that creates time with your team. You'll learn more about the routine piece in my later chapter on TMM Principle for FOCUS
- Create a communication plan for your team. Ensure your routine has both the buy-in from you as well as your teams on how and what should be communicated. A communication routine and channel that works for you may not be the one that works for your team.
- Recognition Plan – Assign a member of your team to help develop recognition strategies that give your teams a sense of appreciation for their contributions
- Performance Metrics – Make sure your teams understand the metrics and what you are expecting. Then determine the frequency of how performance will be shared. I like using sports because I find it's easily relatable. Most coaches will tell you some of the key stats you need to obtain for an efficient game. Best way to coach is to share the stats whether good or bad so the player understands where they need to continue to improve. It works!
- Schedule 1-on-1 sessions (I call them "Ignite Time"). Determine the frequency whether it is weekly, ,

monthly and schedule the time.. Make sure it is recurring for the year so te time can be accounted for. It's similar to going to the dentist. You have a 1 on 1 dental work session. Your dentist provides you with a recap of what looks good and where you need to focus some attention. They give you the goodie bag and what's next? A small card that tells you your next appointment day and time. As a patient, the dentist has my trust and confidence because I know they want to see me on a frequent basis.

TMM PRINCIPLE # 4:

FUN – LEAD WITH A FRIDAY DNA

I don't know about you. But I love some Fridays. In my younger days before I was married, Friday's for me were straight happy hours after work—drinks, appetizers, laughs with friends and some of my co-workers. We would have a ball. Why? The week was over. All the hard work and commitment to our jobs we could now celebrate. And heck, it may start Thursday evening. No different from today as my daughter will say with a smile, "Daddy, tomorrow is Friday, last day of school. I can't wait!" She obviously wasn't thrilled with school days.

Seriously, from the time I wake up, til getting in my car to start the day, there is just a different feeling on Fridays. I don't hit the snooze button on my alarm. Or if I do, it's just not as many times as Monday – Thursday. The kids are not as grumpy and actually want to listen. Even the commute to the office is different. There is less traffic than normal.

Some companies may even allow a more casual attire on Friday. Not to mention when it's time to get off work, many restaurants may have a 2 for 1 happy hour special. I love Friday!

I attended a social event one Friday after work back in 2016. I was with other professionals in the uptown Charlotte, North Carolina area. We were looking at starting a networking group and, because it was a Friday after work, it ended up being bourbon drink specials, laughs, and venting about work. But the laughs and work stories were what really stood out. Jamison, one of my colleagues, even stated, "Why couldn't work be like this every day?" He shared how there was such a different vibe in the air on Fridays. He felt his boss was not micro-managing as much because she was pressed about getting her own work done.

Conversations appeared to be more personal versus a quick fly-by hello. More importantly, there was laughter in the air. We all agreed with Jamison. As I reminisced on this particular social outing, it simply reiterated why it's important to have fun in the workplace.

The point was that the Friday experience in my career seemed to always be exciting. Yes, exciting. I said it. There was certainly more energy on Friday versus the other days of the week. Maybe it was the weekend activities that were coming up that drove the energy in the office. The point here really is that we need to create a "Fun" environment within our teams. Bring that Friday feel to the office, sports gym, you name it. I don't mean work should be a recess session where you run around all day but I do think a level of creative fun and energy is absolutely required.

My son even taught me as a youth basketball coach to lighten it up a little. During my practice routines I tried to bring aggressive grit to toughen up the players. But I lacked fun. One day my son suggested playing a shooting game called Knock Out. Essentially it's a game where all the players shoot from the same spot and you have a chance to knock the other player out by making shots before they do. The kids loved it. Personally it was silly to me. But it motivated the kids and they were ready to then jump into the aggressive practice mode. Got to love our new generation of youth! ☺

Friday DNA Started for me back in 2001

It was a Wednesday evening 17 years ago in Norfolk, VA and I was the second shift supervisor in a call center with my co- lead Thomas. We actually called him TW. Attending high school together we had our fair share of battles playing 2-on-2 basketball after school in 8th and 9th grade. He had his cousin and I had my best friend, Bruce, and we played at Indian River Recreation center. TW, you know we used to kick ya'll butt! ☺

But here we are, it's roughly 8:00pm on a Wednesday night and I'm at my cube strategizing the call support plan for the remaining month. We managed a high intensity sales team. There were over 100 teammates on our side of the floor at this time of night; however, there was a quiet feel in the air. If it was quiet, then it meant our teams were not driving sales talking to customers. If we are not talking to customers, then we are not making money. This was a call center so there should always be the sound of voices talking to customers.

I was side-tracked only thinking about our month end strategy so I wasn't focused on the team's productivity at all. I was planning and focused on the plan for the following month. So, my attention was not on our productivity as it should have been. All of a sudden, I heard some music being played, Frankie Beverly "Before I let go!" TW is walking around the floor with a small radio playing this song, chanting, "Okay, everybody, it's Ol Skool Wednesday." He yelled out. "Let's get excited and finish this last hour before we let go!" The objective was to finish the evening strong so we could stay on track for our monthly goal.

TW added a little dance in his pace as he walked the floor. I quickly popped up, thinking this wasn't appropriate. I was more of a black and white supervisor at the time. I did everything by the book. TW brought creativity to our teams. So as he walked around, I noticed the teams were stimulated and awakened with a motivated spirit. Some began to shake their shoulders and you started seeing ladies doing a 2-step at the cube. The teams were motivated. The teams were burnt out as we had been pushing all month, but now they were getting reawakened. They needed some FUN!

TW brought a creative energy that was needed to drive production. The next day I pulled our production report for the 8:00-10:00 pm time period. We had our highest productivity in sales over the 2-hour window compared to our prior six months. We did it!

TW created what then became known as "Old Skool"

Wednesday. Teammates could request their favorite 70s and 80s song to be played around 8:00pm. It was fun, catchy, energetic and it drove productivity. I even stepped out of my uptight comfort zone and would dance and sing. And after that, we hit the night club! Yeah, we'd go to the night club on a work night. Those were the younger days.

This kicked off a "fever" of amazing production levels exceeding monthly targets where our unit would become the highest production area across the department of 400 employees. Eventually we shared our best practices with other managers on the floor and it went viral. We had nerf football and basketball competition, contests, after work activities, and amazing Gala celebrations. Our employees loved coming to work and we, in turn, loved leading our teams. We remained demanding leaders, yet knew how to reward their performance.

This energy TW brought to our office I would never forget. I learned and carried it with me. When you are leading people, leading teams, winning teams, it's important to ingest some type of creative fun within your team environment. Do this at work, your sports team, your church. I don't care. Just have some damn fun at work.

Today, TW is a successful entrepreneur as a Real Estate investor and co- owner of Great White Water Sports, a thriving jet ski business near Va Beach, Virginia. When I think of my experience with jet skis, it's fun and exhilarating. This certainly connects well with my friend TW's personality and natural ability of connecting with people. Albert Einstein wrote, "The creativity of intelligence is having fun!" Way to go TW, you were wise beyond your years! Thank you.

Me and Jordan riding jet ski's at "Great White Water Sports" in Norfolk, VA having FUN!

Getting Your Teams Involved

One of the more memorable activities that I ever participated in was a Hoop it Up March Madness campaign in 2009. Why was this the most memorable? Well, because I love basketball! 😊 March Madness is the college basketball playoff event held yearly. Me and my partner, Jennifer, at the time wanted to empower our team to create and drive their own campaign.

The campaign committee generated a level of fun and creativity that far exceeded my initial expectations. They cut out orange colored basketballs for the teams to earn for various metric identifiers. They brought in a basketball hoop for teammates to shoot baskets throughout the week to keep the energy and engagement going. They even created a full court basketball sweet 16 bracket, for the teams to compete against each other.

The point here is that the three teammates we asked to

lead this effort, took a task and allowed their creative talents to enliven our department. Our performance sky-rocketed during the event that really set the tone for continued great performance for the remainder of the year. The team was simply having FUN!

My son shared with me about one of his teachers who was really big into health and wellness. I had the chance to witness this as a volunteer at their school. The students were doing push-ups in class. It was amazing. They had a routine where when it was time to stimulate the brain for learning, there was a routine of various exercises to pump their blood flow. The class loved it! I had never seen this in the classroom. The kids were excited and ready for the day. By the way, she could do more push ups than me! LOL

I was reading this article about Kevin Durant and his decision to play with the Golden State Warriors. For some of you, you may recall Kevin Durant left Oklahoma City to play with the Golden State Warriors. He received a lot of criticism for his decision. I didn't understand it either. If you follow basketball, you'll recall Golden State already was a stacked team with all-star talent. Steph Curry, Klay Thompson, and Draymond Green. But I read an article in which he talked about the experience of playing with Golden State. Kevin said the following: "One of the joys of playing with this team is that we work hard, but we have fun playing the game. Steve Kerr creates this environment for us!" Kevin Durant and Golden State warriors went on to win two championship titles. Enough said. People want to be at a workplace that promotes fun!

TMM – Time to Ignite Fun – "Lead with a Friday DNA"

- If you work in a business casual office, at times make a jean day on a Monday or any day of the week. Think of the morale when a team knows they can come to work on a Monday with a pair of jeans on. "I'm just saying…"
- Incorporate some fun and easy games that are not too distracting. Bingo has always been a big hit in my experience.
- Create a "Show Your Team Love" day for your office. Spoil the heck out of your teams. Cook breakfast, buy lunch, and do their job for an hour or two if you can. Just make sure you're a good cook. And if you do their job, don't mess up their numbers.
- Team builder social events outside of work – Teams want to get out of the office. If your budget allows for quarterly fun activities, spend the money. Shout out to my wife who put together an AWESOME team builder event at laser tag to kick off our 13U AAU season. The kids loved it and we had a great season. Go Team!
- Leverage sports to help build themes in the office i.e. Super Bowl, World Series, NBA Finals, Soccer World Cup. This seems to always be a big hit. Check out Linked in, and you'll see some pretty cool office pictures. I even see churches and restaurants get really involved wearing sports jerseys for these huge sporting events

- Community Volunteer Events – Participate in a volunteer event outside of work. Giving back puts a smile on faces and brings team collaboration
- Stay Healthy – Do some type of walk or diet challenge. I'm apart of a running club in charlotte. Many talk about their walk and healthy eating campaigns they participate at work.
- Keep it simple – Be like my good friend, TW. Turn on some music and encourage some dancing and laughs. I promise your teams will have a ball with it.
- Bowling and golf events are always a hit. Listen, whether you can bowl or golf, these types of venues just bring some fun. It's not hard work. Whether you bowl a strike or a gutter ball you can always turn around, walk back to your seat and have a cocktail. Only one! 😊

PRELUDE INTO THE NEXT SET OF PRINCIPLES

Congratulations. You have made it through the first four TMM principles. How do you feel? When we think of the frustration that goes into leading teams, I'm hoping these principles get us off to a good start. I tried to balance both personal life and career experiences so the messaging makes sense.

Here's the why and the purpose of the first four principles. Based on my experience I believe if you set the tone with these principles, you have won the trust of your people. The trust of your people has to be #1. I'm using a lot of fatherhood and office-related examples. However the principle may apply in many professions. Many of us have kids or will soon, so you'll be able to relate, as leadership takes place both on and off the court as my old coach used to tell me. Basically, if you work in a people-serving environment you have to win their trust if you want them to perform well.

You will notice, I did not talk about how to do process innovation or how to manage a financial budget. This was not research but my experience. This is all people-driven where you need to win their trust, morale, and confidence before you can actually lead them. Think about it, we've established a solid **Foundation** that your team can stand on. We've discussed **Finding the** *talent* within you and the team. We hit the **Fundamentals** understanding there are basic routines we must get right. Then we're saying we're going to have some **Fun** along the way as you lead your team in achieving results. I don't know about you but if my boss, coach, principal started out here, he or she would have my full loyalty and support. I would give them 110% effort always.

Let's keep Going!

TMM PRINCIPLE #5:

FLEXIBILITY – LEADERS CAN'T WEAR THE SAME PANTS EVERY DAY

Here I am sitting in my creative book writing class on a Saturday morning led by Pulitzer Prize winning journalist Glenn Proctor. Shout out to Glenn who was my book writing coach. He kicked my ass every day, as he would say. He's a tough one to please but he put faith in me that I could finish this book. Thanks, Coach!

Back to the story. We had to write about giving up some piece of clothing that we may have worn or kept for years that was meaningful and we did not want to let go. I wrote about a favorite pair of jeans I used to wear in my early 30s. They were a pair of Sean John Jeans made by Puff Daddy's clothing line. I thought I had some cool swag when I wore these jeans. Swag meant you had a style that illustrated confidence and appeal. This was an Ol Skool term now but you follow me.

These were the jeans I would wear when hanging out with my friends going to a night club or nice restaurant. With a cool belt and a designer shirt, let's just say I was ready for the city. What made it even better, my wife bought me those jeans. My wife has that natural fashion eye for nice clothes. So, if she bought them, I'm looking good. And when I wore the jeans all I thought about was how cool I looked, especially if I was rocking a nice button down shirt to go with it.

About seven years later, I was still wearing these jeans. My wife would tell me, "Why are you still wearing those same old jeans?"

I thought "Wait, you bought me the jeans so why not?" But she was right, the style had changed and plus I'd gained a couple of pounds. Well, maybe a few! Baggy jeans started to play out. We then went shopping and I bought a pair of nice straight leg polo jeans. As much as I liked my baggy jeans. The straight leg felt good and comfortable. Instead of a cool swag, I felt more like a man on a mission looking to conquer the world. I wanted greater success. Especially as my son got a little older, my goals and dreams were changing. The jeans represented my shift in mindset. It was time to change the game. I needed to be flexible letting go of the old and going after the new. DANG, I miss those jeans! LOL

My first leadership lesson in becoming Flexible to Change

My first BIG experience in having to be flexible to change was back in 2007. I had made a career move, now working for a very large Fortune 500 company as a loan officer. This was a huge transition for me as my career to that point was really serving in a leadership capacity. I was approached by a former colleague of mine regarding the opportunity. But it was more of a sales job in my opinion and I like the comfort of knowing my paycheck amounts on the 15th and the 30th. LOL. I went after this move mainly for the money. When I was told I could make over $100,000 per year processing loans and I didn't have to manage people. I thought damn, why not?

This was a different change. Number one, I sucked at calling clients trying to get them to refinance. As much as I like to talk, I was horrible at my talk off. I had no personalization. I was going straight after the loan application. Getting denied. Long story short–I did get the hang of it and started banking these loans in. The money was exactly what my colleague said. My family started traveling and getting our shop on. Life was good and pleasant. But it all changed. After about six months, the mortgage crisis hit. Shucks! I went from 10K checks to less than 1K in a snap of the finger. This was a quick lesson, by the way, to not chase career moves solely over money!

I had to get a new job to put food on the table so I got hired on as a manager at a local call center. This was a different change for me and I had to be *flexible*. This was not a Fortune 500 company. The structure was something I wasn't used to and most of the employees were all new to the contact center business and this may have been their first job. These new hires were literally between 19-24 years old. I'm like, are you serious? And they wanted me to manage this group because of my experience. I'm like, you cannot be serious!

This was a tough transition. I managed by human resource rules and policies. For excessive absences and performance issues, I managed this very aggressively to either manage you up or out. This group immediately had just that. Absenteeism and performance issues. I was quick to write folks up and look to terminate. I was then pulled into the office by the executive. Basically, the message was I needed to loosen up and manage with more empathy and not the termination stick. My thought. Heck no. You show up on time and you perform. If you can't, you can't work for me! And if the office could not support my leadership style then I may have to leave. But, and I do mean but, I needed that paycheck to put food on the table.

Shucks. Time to be more empathetic. So, when folks were late and having issues I became more of a father/mentor. Helping this team understand the why behind the impact of their absences and tardy's. And why performing at a high level was so important. Making them understand why I coach the way that I do.

It was tough for about 60 days but they finally got it. I became more empathetic and understanding regarding some of the personal issues causing them to be late. No ride to work, babysitter issues, working two jobs. But they raised up their level of performance. They saw how much more money could be made if we achieved certain collection goals and they were intrigued and motivated. They wanted money. In short, out of 31 teams, we became the #2 team consistently over a 6 month period. We were kicking ass!

They were making money. I was making money. But it took me becoming *flexible* in my leadership style to get there. We did it. I had to remove those old jeans of mine and sport my new ones. I had to grasp and be flexible for change.

So from that moment on, I took on a new mindset realizing flexibility for change was critical. If I showed flexibility with an understanding around the why, the teams would rally, not behind me, but with me. More important, change can be embraced when communicated with specific details including the "Why" behind the messaging.

I took a personal oath to ensure I understand the changes so I can be a team player and effective leader if illustrating the change and the benefit. I see this so often today with teams. Especially within this rapid-paced environment we are in. We move so dang on quickly we don't ask why and we become stubborn on change because we simply don't understand. Our egos are oversized and we think we have all the answers. Or, we are intimidated by how changes will impact our ability to lead. That's not how this works. Typically, changes are meant to improve.

Learning the Neuroscience of Change

During this period of running my consultant business, I later discovered an interest in learning about the neuroscience of change. Between attending webinars through the neuroscience institute, online research, listening to audio CDs, I've learned more about the brain aspect of responding to change. Strap on your seatbelt for this next section as this gets a bit technical.

The formal definition of neuroscience is the pathway of data flow within the central nervous system dealing with the chemistry, pharmacology, and pathology which impacts behavior. In simple terms, change ignites a sensitive area within the brain that causes a pull effect. As a result, many are just reacting to a natural alert within the brain that says RESIST. So, the more we as leaders hammer down change and create an "it is what it is" culture, the brain is causing a natural pushback. The brain is needing a level of cultivation of empathy and understanding through effective communication, empathetic statements, and ongoing leadership coaching to help your teams respond favorably to change. This was a masterful learning experience for me in how I would have shared change and become a better communicator of change. DANG, I feel smart! ☺

As a certified leadership trainer, I teach my courses on "change leadership." I really illustrate this piece and it has become a huge "Ah ha" moment for managers attending. I think about my messages in the past wrapped around changes. I relied mainly on my charisma and likability for my teams to trust me as their leader. However, maybe the science aspect would have allowed me to continue to coach more with empathy and probe more to ensure my teams fully understood changes and agreed to accept and move forward.

Teaching change leadership has been by far my more enjoyable course to teach and continue to learn. With the changes happening in our world today, if we are not flexible to what's changing around us and we are going to fall behind. So, I wanted to make sure I had a better nurturing experience when coaching leaders in being flexible to model change.

This also helped me to become more empathetic to those who struggle in adapting to change, especially as leaders. Maybe because I experienced my hiccup sooner rather than later in failing to be flexible to change, I became a sponsor of change early. So when I heard leaders voice their frustration for change I did not how to coach them effectively through it. I took more of an approach to keep away from what I perceived to be negative thinkers and tagged them as resistors of change. Man, I wish I could have some of those frustrating moments and conversations back as I would have been a better coach in leading them through change.

Personal Story picking on the Baby Boomer Generation

A fun example I like to use when I'm speaking on change is the baby boomer adjustment to the flat screen smart phone. I'll share the example of my father who was a proud owner of a flip phone. The flip phone could easily fit on your belt strap, and only twelve buttons to push. My father held on to this flip phone until 2016, when he finally decided to go with a smart phone. He did not see the value of all the afforded information that he was missing through his standard flip phone. His mindset saw confusion with the many applications, too big with less comfort, and information that he felt would not be needed. He was a resistor to change.

After a few years of me and my younger brother, Dexter, getting on him about the phone, he finally made the shift. I look back at joking with my father for sticking to the flip phone so long. But then again, I failed to really explain the "why" around the change. I never gave him the specific benefits of the Smart Phone. I could have taken the time to sit down and walk through some of the more user friendly applications that he could take advantage of. My dad is a natural bass guitar player. An AWESOME guitar player by the way. There were several applications through which he could have been downloading music and guitar playing information. All I would say was, "Dad, it's time to keep up with the times and convert to the Smartphone." Basically it was my way of saying, "Dad, it is what it is!" But I never really shared the benefit to the change so I was a bad coach educating my father.

Fortunately today, my father is a proud owner and user of the smart phone, exercising the capabilities of GPS, Email, voice texting, and music. He's now with the today's time of cell phone usage. Way to go, Dad!

The example with my father is what we do often as leaders. We implement a change and say here you go. Or, in my father's case I only shared that it was time to change. We roll out changes with no rationale behind the change. We don't take the time to discuss the benefits or allow questions to be asked. We act like it's just a new change that falls from the sky and your people are expected to execute. That's not right and some of your great talent may leave you because change is going to be perceived negatively.

Change starts with a Growth Mindset – It's all coming together now!

Here's what really hit me about being flexible for change. When I was preparing to teach a course on Growth, something really stood out. It hit me. Many of us don't have a plan for growth. So we naturally resist change. If our teams all had a plan for growth it would make more sense and become more acceptable because you have a well-designed map. And you appreciate your leader having a plan for growth which will require some level of change. AH-HA!

Simple Fix Leaders. Before you start driving change, make sure you invest a plan for your team collectively and individually that encourages growth. If they have a personal plan for change, they are then automatically injected into a change-driven culture. You have a plan and your team has a plan. Wow, this is not that complicated!

Getting the audience pumped up for Change!

I was teaching this class at a local business expo at Piedmont Community College in February of 2017, I was getting the audience all fired up for change. I would make them repeat after me, "Success!" "Begins with Growth!" "Be Intentional!" "First I have to Change!" This was a 30-minute segment all about personal change. I wasn't teaching about why they needed to change for their company. But the why then needed to change for their own personal development. Throughout the class, I would make them say it repeatedly so it could stick to their hearts. "Success!" "Begins With

Growth!" "Be Intentional!" "First I have to Change!" I had a lady confess to me after the class that she was one of those resistors of change at work. And for the first time, she realized she needed a plan for change and we started a coaching relationship at that point.

As a leader, your ability to be flexible for change is crucial. The amount of change we will see over the next 10-15 years will be breathtaking and unbelievable due to technology. Your ability to first accept change as a way of life for both yourself and others will be extremely beneficial to the ongoing growth of your teams. As the wave of Artificial Intelligence sweeps its way across the globe, those who are not flexible for change will fall behind. This was my dad with the old flip phone. Ha, ha!

In November 2017, CNBC reported that robots may replace 800 million workers by 2030. What will enable job stability are people with excellent soft skills and emotional understanding. This is why I'm pressing the topic of being flexible to change. This is why I'm so passionate about this book and the soft skills that I'm sharing. It doesn't matter what industry of work you are in. Hiring great leaders will always be of high demand despite the expected shift with automation. And I believe this is you!

The TMM principle of flexibility may be the more critical of those you are reading about throughout the book. My early years of leadership were about coaching and developing others based on their job skills and requirements. Today, I believe leading change through a flexible mindset is going to be a key requirement of leadership over the coming years. The world needs strong leaders of change.

Look at what's happening.

1. NFL running backs are no longer getting drafted early. Teams want a pass heavy offense.
2. The NBA is now played from the perimeter versus the inside. By the way, does anyone even care about the dunk contest anymore?
3. Instead of going out to eat, simply call Uber Eats and have it delivered.
4. Kids don't ride bikes at the same rate as they used to. Now they stay inside and do playstation and games on the IPAD.
5. My wife and I were looking at cars the other day in the car lot. No prices were on the car windows. We had to scan to get the car price details. I'm going up to every car having to pull my phone out to scan for details. SMH.
6. My son says he doesn't have to drive when he's 16, he can just UBER. He'd rather be on his phone while riding.
7. What about head coaches? Does yelling and threatening players even work anymore? It motivated me back in my day. But these kids are very intelligent; they want logic, reasoning, and empathy. Let's just say they are spoiled! (Ha, ha) So coaches have to change their delivery.

My point. So much change is happening and we have to adjust. We have to be flexible as leaders, understanding the needs and wants are different today. We have to parent differently today. Preaching has to be different. Teaching, etc We simply have to get rid of those 'old jeans' and put on a new way of thinking. Thanks again, Mrs. Mike, for making me get rid of those old jeans and changing my mindset. ☺

TMM Time to Ignite Flexibility

- Develop a growth plan for you personally that encourages change. If you have a plan for you to grow, this will help you understand why change is important for the people that you lead to understand it all.
- Use facts and details supporting why change is critical. Leverage the data in the market for where you work or lead to help explain changes you may be making within your business.
- Create "Why" discussion forums that give your teams a comfortable and open opportunity to ask about the changes of your business. This type of environment allows candid dialogue and it gives your teams a sense of empowerment to freely express themselves. More importantly this makes you look good.
- Enroll in change leadership courses. The methodology of leading change is critical, in that it helps you understand the questions to ask when a part of change initiatives. You don't have to enroll at Harvard leadership school. I facilitate a heck of a class on change leadership. Just email me. ☺

TMM PRINCIPLE #6:

FAILURE – CREATE A FAIL TO LEARN ENVIRONMENT

Baseball includes a good concept to fail and learn. We've all heard of the saying, "Three strikes and you are out!" This originated through the game of baseball. I played a little baseball and I know all about facing those three strikes. I've been struck out a few times in my little league career for sure.

Batters have an opportunity to make contact with the ball through a swing at the plate. The goal is to get a base hit or homerun. If the ball is pitched within the strike zone and the batter does not swing or swings and misses, this results in a strike. What I appreciate about baseball, is that you have a "real time" learning experience as you stand at the plate. Maybe you did not swing at the first strike. The next pitch you fouled right. You then fouled left. The ball is pitched high away from the strike zone and you were patient enough not to swing the bat. Finally, the next pitch you timed your swing perfectly and made contact with the barrel of the bat, resulting in a base hit or maybe a home run. The batters' success at the plate is not always predicated by the first or second swing. It's the ability to learn from the pitches thrown while making adjustments to execute a good hit!

My point is we all need to experience a little failure to learn. Within your day job, you shouldn't be so concerned with failure and missing the first pitch that you fail to grow. I've seen managers in my experience who never even make it out of the dugout because of their fear of failure. They were simply doers and not thinkers. I've watched talented players on the football field never show their full potential. I know high school quarterbacks with a great arm but scared to throw timing routes. They'll only throw if the receiver is wide open. Well, colleges overlooked these quarterbacks because these are the throws they will have to make in college. The list goes on. Whether you are trying to teach new material, experiencing new relationships, you may have to fail to become great at your craft. If you don't fail, you'll likely end up in this bubble of safe complacency which will never allow you or your team achieve monumental heights. Trust me, I've failed plenty and I am grateful for those failures as I learned so much.

A personal wake-up call!

I conducted a training session for a local middle school in Union County outside of Charlotte NC, on the topic of rebounding from failure to success. As I was piecing the slides together, I wanted to ensure the message was one with which the students could relate.

There were a few major superstars I used that they could easily identify. My thinking was that trying to encourage over 100 students that failure was acceptable, it needed to be catchy and interesting. I chose Michael Jordan, Oprah Winfrey, and Bill Gates. What was intriguing after I did my research was their journey to success. The common experience for each was that they all had been up to bat at some point with a swing and a miss. We often become so captivated by success we see on television or what we read that we often overlook the journey. But more importantly, through their learnings they each were able to hit homeruns.

Bill Gates – Bill Gates was a Harvard University dropout and saw his first business fail at age 17. I was impressed that Bill had that level of entrepreneurial spirit at such an early age. Then to have the intentional mindset to again start his next venture that later became Microsoft. Bill became the youngest billionaire at age 31.

Oprah Winfrey – Oprah was dropped from an ABC affiliate news station in Baltimore, Maryland. The public eye did not appeal to Oprah. She was part of a new show that did not perform successfully. Oprah was to blame for the failing show. But Oprah went on to have her own afternoon talk show and the rest is history.

Michael Jordan is regarded as arguably the greatest basketball player in the history of the sport. He's been vocal about his story about failing to win a championship early in his career. He talks about the playoff frustrations with the Boston Celtics and Detroit Pistons. The physical beating he took. However, it elevated his thinking and desire to get stronger. Well, we know the rest of the story, as Michael won six championships and never lost in the finals.

What's most important about Bill Gates, Oprah Winfrey, and Michael Jordan was that they each experienced failure. The critical piece of their unique stories was their ability to overcome obstacles in reaching their height of success. I reminded the students that learning would allow them to prosper as individuals though they would never reach perfection. However, when I asked the students who felt pressured about trying to be perfect, nearly everyone raised their hand. Something wasn't right. I wanted to dig deep to better understand why and self-reflect. My son was in this same middle school so I got to thinking that Jordan likely felt the same way.

Dad needed to be put in check!

One of my favorite books, is "Good to Great" written by Jim Collins. It is about a concept of developing excellence through a methodology that moves you from good to great. I often would quote to Jordan about excellence in academics and sports participation: "Jordan, there is no such thing as being good, but you should strive to be GREAT!" Maybe that's just a bit extreme for a middle school student to handle. I was always pushing for excellence but I never covered how excellence was obtained. I never explained that excellence comes with learning from failure.

After I spoke to the school, I immediately wanted to talk to Jordan. When I picked him up from middle school basketball practice, I asked the question. "Son, do you feel pressured by dad to not make a mistake?" "Do you feel pressured to be perfect?" His answer was "Yes" and "Yes!"

His response reminded me of why leaders and their teams often worry about making tough business decisions. Immediately it made me realize why at times our teams won't take risks. They were concerned about the failure and those repercussions. I would tell my mentees all the time, "Your teams should be the CEO of their desks." "Make the tough business decisions." But yet, I would push so hard for perfection. It makes sense why I could not get some teams in the past to think outside of the box and take risks. As their leader, I was pushing for an error-free environment with high production levels. My mistake was that I may have been pushing too hard, constantly showing the numbers and talking about why we needed to be number one. Who would feel okay about accepting imperfection if all I talked about was the pursuit of perfection. Like my son, I realized at work and at home I was not creating an environment where my teams could really feel empowered to take risks and learn. I was not creating what I now call a "fail to learn" culture.

Truly I was not being an advocate of taking a swing and a miss. Yet attempting to drive perfection through a no-mistake tolerance versus encouraging mistakes as a means of learning and growth. Thus, we could coach on how to see improvements in the future. I was supposed to be a good coach, providing empathy and support by helping to design a plan based on a failure that would ultimately lead to greater success. Funny thing. Heck, I'm far from perfection, so who was I to raise this somewhat of an unfair standard!

I remember driving home that day with Jordan, apologizing for the overly high standard. I explained the "Why" behind my thinking in that I wanted him to be successful. But I understood that his success would come with inevitable mistakes. As long as I was shielding his thinking and creativity, I would be disrupting his plan to develop as a young kid. "My apologies, Jordan, and thanks for forgiving me!"

Back to Work - Building a Fail to Learn Environment

Heard a great idea from Brian, one of my colleagues working in the manufacturing business. Brian explained that he used failed opportunities as a means of celebration versus a failed demotivating experience. Brian incorporated what he called, "Guess what I found out?" news of the week. If someone within his team had a failure within their assembly online such as a quality defect, he would communicate the fix or reminder as a "guess what I found" topic for the week to be shared as general information. Or if he simply discovered information that was relevant among his team, it was sent through this communication method. It quickly became contagious. Failure later became a non-embarrassing experience. Dang, a pretty smart approach. Nice job, Brian. I need to use that one!

Sharing some of my personal failure! A Big Mistake I made when I started my Coaching Business

From my own experience, starting your own business is certainly different than working in corporate America. I realized that I was horrible at quite a few things.

Strike 1 – Networking was not my strength. I would go to networking events selling myself as soon as I met someone shaking their hand. "Hi, my name is Richard Mike, owner of The RJM Experience. I have a leadership coaching business specializing in developing leaders of tomorrow." This would go on and on. No wonder why I lost many business deals. A meeting with a member of my local chamber in Matthews, North Carolina taught me a lesson. I was selling before I even built a relationship. I wasn't asking enough questions about the person I was meeting. Having no clue if my business was a good fit for them because I never gave them an opportunity. Chuckling now as I recall those conversations. I was horrible! But the failures have helped me today, where I've learned to connect with people in a genuine fashion. This was a good lesson learned and has helped me in my current profession and community.

Strike 2 - My marketing was bad. This was not my thing and I had a hard time marketing myself on social media. My posts were rather boring and I was told seemed scripted. My video posts seemed very rehearsed. My son used to record my videos through the phone and we would re-record up to 10 times. Even Jordan said, "Dad, just be yourself and relax and talk!" He was right. I wasn't authentic. If my potential client saw the video and thought the same, no one was going to do business with me.

Strike 3 – There was a song by rap artist, Drake, titled "In my feelings!" If a prospect told me "No," I was in my feelings, meaning I would take it personally. I had an ego. I wasn't cool with the rejection thing. Plus, I felt I was pretty darn good at what I do. They were not going to hear from me again.

Then I went through a particular sales training class and I read this book *The Entrepreneur Mind* by Kevin Johnson. Awesome book on strategies to help you as an entrepreneur. I later realized that I did not have a follow up routine to reach back to my prospects. While I was taking the "no" personally, it may have simply been bad timing. I think about the guy who sets up my home security systems in our neighborhood. Folks would tell him no and he would continue to sell security systems in our neighborhood. He would just come back a month or two later. Well, he ended up installing quite a few security systems on our street. He didn't take the prior no's personally and continued to follow up in the event that we might change our mind. He wasn't into his feelings like me!

Your leadership responsibility I believe is helping to create a "Fail to Learn" mindset. Of course, we want the best effort, but we must absorb the failures as the biggest asset to learning. You did not get the job because you fell short on the interview. You failed to make your goal for the month. Your boss rejected your business proposal for an increased budget. The woman you had a crush on for years said she's not interested. No need to get frustrated. These are all opportunities to learn from a failure because you now embrace these as opportunities to grow.

Final note on failure. If you want to simply be inspired on why failure is important, I will reinforce my message within this chapter. Please do a Youtube search on Will Smith Failure, "Fail Early, Fail Often, Fail Forward!" This 2-minute message invites a level of comfort regarding failure. I'm EXCITED now as I'm playing this video and I'm writing. In short, successful people are a result of their failures. Allow your teams to fail if you want a successful team.

Why is this chapter so important? Here's the deal. I say all this in an environment where we as people want everything now. Honestly we are under the pressure every day to get it right. Coaches, managers, principals, pastors, all of the leadership roles that are out there, you are under the pressure to get it right and it get it right quickly. You may think 'how the heck can I invite failure into my team?' Set the expectation early that you have a plan for success that may include timing. I never tell a boss when I'm interviewing for a role that I'm going to come in and drive results within my first 30 days. I showcase the fact that I will have a plan for success and within my first 30-60 days I will be documenting how we will get there. This plan will include learning experiences. This allows room to cultivate and understand where failure may reside so within our long term plan we have realistic target dates that will allow us to get there.

Most head coaches do not take on a job and say I will win the championship their first year. That's the goal but they showcase the plan on how they intend to get there.

This allows room for failure and learning to occur on the path of your successful journey.

TMM Time To Ignite a Fail to Learn to Culture

- Share a message with your team, acknowledging that failure is going to happen and you will create an environment where you learn by mistakes. No threatening action. If so, you will not see growth within your team!
- Create a "Guess what I found out?" library that houses helpful tips for future reference. Note: use any creative name that you like. The point is that you have some mechanism of sharing learning opportunities within the team.
- Lead by example. Don't try to be perfect in the eyes of your team because there is no such perfection. Share areas where you have made mistakes so they can feel comfortable that it's happening among everyone.
- Pilot all new projects. Don't jump out the gate assuming this is the final product. Set the expectation of the pilot so your audience understands there may be errors or changes needed. This phase sets the expectations that failures will occur and will be used for improvements. My business could have benefited from a pilot launch for the first 90 days with inspection points weekly to measure the success. I waited until after six months to figure out why my expected results were not happening.

TMM PRINCIPLE #7:

FACILITATE – GOOD LEADERS FACILITATE THEIR OFFENSE!

My wife and I were organizing our son's 7th birthday party in Seattle, Washington. My wife did a great job preparing for his party. It was held at a local pool facility. The kids had a blast playing water sports, jumping around, and eating pizza. After two hours it just seemed there needed to be more time to enjoy Jordan's party. I believe it was really me that felt there needed to be more. I'm the party animal of our family.

As we came back home, I decided to have a random block party right outside of the house. This was for the kids and adults. Now, you have to understand, living in Seattle you do not get many days where it's not raining so you have to take advantage of a clear day. We got the grill out, parents brought over food, and my neighbor Dave pulled out his DJ set. Dave loved music and I was amazed that he was setting up his turntables and laptop in my garage for this neighborhood party.

We had a blast between the kids playing games and the adults all laughing, eating good food, socializing, and dancing as Dave was rocking some ol skool Run DMC and Dougie Fresh. Of course me, the host, had to get on the mic and make announcements throughout the event to keep things organized. Then my neighbor and friend, Carlos, yelled out, "Okay, Mr. Facilitator" we got you covered. Carlos came over and said, "Man, you can throw a party and you do a great job facilitating to make things happen." I think he was really telling me that I liked being in control. LOL!

When I think of Carlos, I always think of him being the first one to call me facilitator. I thought about it and he was right. I really appreciated Carlos and his family. Carlos was an encouraging friend with a great family. Carlos was actually the first one I talked to about maybe starting a business one day and writing a book to help young leaders.

He really helped us out when our daughter was being born, taking care of Jordan for us. We really appreciate you, Carlos!

Being a Natural facilitator

I played basketball as a point guard when growing up. I naturally enjoyed leading a team and helping to make others better. My brother, Dexter, even told me one day for my birthday that he felt I always did a good job of making others feel valued and special. That was pretty cool. So maybe I truly knew how to facilitate to get things done. I really just wanted to put my best foot forward in helping people create experiences that were memorable. More importantly I wanted to see others reach their full potential.

It hit me. Good leaders must be able to run an offense like a point guard or quarterback and facilitate to make their teams better. They can't do it all. I was recently reading the Bible in the book of Exodus. Moses' father in law told Moses, "You are doing too much." Moses needed to delegate or be a facilitator in using the talent around him to execute God's plan among God's people. I'll let the pastors read this and take this message away on a Sunday service! 😊

Today's world needs leaders who can facilitate

Recognizing our world has created so many demands among leaders, we often find ourselves taking on a heavy load of tasks or responsibilities that are not always necessary. In management I've seen so many people get caught up in saying "yes" so often their plates get completely full. On top of their daily demands in developing people, running a team or organization, they end up working these long and exhausting hours. I was just meeting with a client who was looking to create better structure and routines for the business. They owned a small insurance firm out of Norfolk, Virginia. The owner seemed to find himself doing administrative duties, quality review, contract negotiations, and several more tasks to where he was truly burnt out. His feedback to me was that he just felt like he needed to do so much for his office to remain viable in the market.

We talked about he and his team taking the strength matrix to help identify his team's strengths. But also, we wanted to find out whether or not he was delegating. And not just delegating by sending work to one person and then to another. Or loading one person with all the "to do" work. But could he effectively facilitate his office to maximize the strength of his team? This goes back to the Find your Talent principle. He was the business owner but he was spending way too much time in his business because he did not know how to facilitate.

Could you imagine Tom Brady, seven time Superbowl winner and five time Superbowl MVP not being a facilitator. I wonder how many championships Tom would have won if he tried to win games by himself. But what Tom has done well over the years is facilitate his offense. With so many defensive schemes and various personnel that he sees each Sunday, he has to identify ways to facilitate the offense to help make others better. He must know the role of his tight end, receiver, running back, and linemen, and make adjustments for his offensive line to block effectively. Tom Brady is a facilitator!

Magic Johnson, former point guard and five time NBA Finals champion with the Los Angeles Lakers, was known throughout his career as a great facilitator. As a point guard, I remember as a kid and now as an adult watching youtube highlights of his amazing ability to pass the ball. Magic had a unique ability to see the floor, locating open players for a jazzy pass before defenders had a chance to react. At 6'9" as a point guard he could have dominated by scoring the ball on each and every position. But Magic understood his role as a point guard that he had to facilitate to maximize the talent on his team to win. Magic Johnson is a big time entrepreneur and has put together a successful business resume through Magic Johnson Enterprises, owning movie theatres, Burger King and Starbuck franchises, and having thirteen 24-hour fitness gyms. What a great facilitator he is!

What's important about being a great facilitator is that you have a team around you. Facilitators do not work in silos. Facilitators work in a people environment, understanding the importance of leveraging the talent to be successful. Tom Brady and Magic Johnson may both have a unique talent for their respective sport. However, their ability to lead and facilitate responsibilities among their teams has earned them success, not their individual skill alone.

"I feel like I can't get ahead"

This is a feeling I believe we all go through as leaders at one time or another. The demands have piled up from your boss, customers, partners, and your teams and you feel like you can't get ahead. You are frustrated. You may even have said, "Forget it!" Now, it's time to slow down and think and gain control of your day. This was a mistake I made several times in my career. My personality is just wired to say "yes." Being one who likes to support others and having a genuine heart, I say "yes" thinking I'll come up with a way to get it done.

Too many times I've over-committed to my wife and kids, making promises that I failed to follow through. "Daddy, can we go to chuck-E Cheese?" "Yes." "Dad, when you come home from work can we go play ball at the gym, "Yes." "Honey," my wife would say, "can you cook on the grill tonight?" "Yes."

In many of these instances I'm tired. Yet, I say yes to try to please them each. But then I cause disappointment if I fail to follow through. The truth is that as hyper as I am and as wired with energy, I do get tired. I just want to come home and relax. Sometimes I follow through but I'm burning my own body out. Or I leave them in disappointment for breaking a promise.

Here's what I want you to ask yourself to begin making that shift as a facilitator:

- "How can I delegate some of the work that is on my plate?"
- "What skill set stands out among my team or people around me which would allow me to effectively assign new work?" Remember the TMM Talent Matrix in principle 2. Use it.
- "What new champion roles can I now give?"
- "Am I taking on too much and if so…where can I facilitate the work with others?"

You certainly want a winning team at work to help you exceed in delivering for your clients and organization. So, take a time out and incorporate a plan to become an effective facilitator. Instead of starting your day off figuring out how much you need to do, I suggest that you incorporate a plan that determines what you can delegate. Many of you may be working. You are the facilitator!

TMM – Time to ignite the Facilitator in you!

- Keep your "Find your Talent" matrix visible in front of you. This serves as a reminder to use your people to delegate work that is applicable. You don't want to delegate a newsletter communication to someone who is not good at written communication.
- Make sure you plan some time to help determine your workload for the day and who you can facilitate work to. Great facilitators are proactive, recognizing the defensive schemes and avoiding being reactive. Your distractions are often where you need to begin facilitating. Urgent customer escalation. You are double-booked for a meeting. Someone called out for the day.
- Here's a tough one. Know when to say "No Mas!" You can't take no more. Respectively share that you have a full plate of responsibilities that need to be completed before you take on additional work. However, make sure those items you are working on are high on the priority list. Not a bunch of small task items.
- Share your facilitator plan with your team. This is a "Win" for you. Your teams will rally behind you and feel valued knowing they are a part of your execution plan
- Practice at home. Especially for some of you who have some means of "control" like myself. Start assigning more work to your kids, spouse, or call-in service for help. Have the kids to walk and feed the

dog, wash dishes, clean the table and sweep the floor. I use the kitchen because this is where most chaos happens because of it being a high traffic area. No kids or spouse. No problem. Call-in services to complete your laundry or clean your home if your days are quite busy.

TMM PRINCIPLE #8:

FUTURISTIC – INNOVATIVE THINKING KEEPS YOU IN THE GAME!

It was 1985, and I was nearly ten years old. I remember riding with my father to pick up a VHS movie to watch. My dad always enjoyed renting movies back then. For some of our younger generation, VHS movies were these large tapes that would go into a VCR. VCR players and VHS tapes were how you watched movies back then. Unfortunately, there was no Firestick, Redbox, and other streaming devices during that time. Dang, I'm starting to sound old!

My father was really pumped about seeing this movie Terminator. The main character in the movie was former Olympian bodybuilder and California Governor Arnold Schwarzenegger. My father was a big fan of his because he too was a bodybuilder, extremely heavy into weightlifting. The movie depicts Arnold as the terminator who was sent from the future of 2029 to 1984 to stop a robot in his mission to kill a soldier's mother. Toward the end of the movie, there are scenes of robots fighting each other and shooting humans.

I remember asking my father, "Dad, will there really be robots one day?" He said, "Yes, very soon." I thought robots were just for movies and cartoons and I could not see the reality of robots happening in real life. Well, look around us today as we now have robots that have surfaced to help automate work. If you recall in chapter 3, I talked about my dad being a genius. I guess he really did know all the answers.

As I was reaching the end of writing this book, I debated on the number of F words I was going to keep or take out. I contemplated on this one greatly because I didn't feel this was people-based. But I realized the "F" principle of Futuristic was critical in, at minimum, the thinking.

I remember telling a group at a business dinner about eight months ago, how I felt the needs of leadership had not changed over the years or the century. I felt through all the digital advances and technology ever-evolving, the one constant in leadership was the people aspect—leading people with dignity, trust, respect, and empathy. Leaders were those willing to take a stand for what they believed and what their people believed. Leaders would walk forward, always bringing others with them. Leaders were excellent communicators who could motivate and inspire a team. Leaders I felt meant leading people!

But as we are in this innovative climate, leaders need to be futuristic thinkers. This is key. Thinkers may work hand in hand with the inventors or creators of some big time cosmetic change. Simply think outside of the box. Be different.

I've been in the leadership game with companies for over 20 years, but being an innovator was never an expectation. It was always about performance and people development. Most of your bosses do not ask you 'what did you create differently today? What's the new improvement that we are rolling out?'

Typically it's: are your teams executing against the goal? So, if you want to stand out and be effective in driving a winning team, this is one area where you may say, "Richard, this does not sound like soft skills." It may sound like I'm being scientific or something but that's not the case. I'm simply embedding an awareness that winning teams have to think differently in changing the status quo.

I love this basketball example going back to the Golden State Warriors. This was the first championship they won with Steph Curry, Klay Thompson, and Draymond Green in the 2014/2015 season. The Warriors were down in the Finals series 1-2 to Lebron James and the Cleveland Cavaliers. The odds were certainly in the Cavaliers favor. Golden State decided to replace their center Andrew Bogut with Andre Iguodala in the starting lineup and play small ball. This was a very non-traditional line up. Andrew Bogut was the Warriors center at 7'0. Andre Iguodala was a small forward at 6'6. This at the time appeared to be a crazy move. It was different and not well received by the Golden State crowd.

Fast forward, Golden State won three straight games

to eventually win the title. Cleveland could not adapt to the small ball offense. This was magical. It worked. It was futuristic. Why futuristic? Well, the small ball offense has become the norm across basketball among different levels. This is 2021 and you are seeing several teams playing with a non-traditional center to add more athleticism.

But wait, here is the cool part. Steve Kerr, the coach, did not come up with this plan of going small ball. It wasn't the general manager or the president. It wasn't the assistant coaches. It was a 28-year-old member of the Warriors staff that was considered a special assistant. He did video editing, collected playlists, organizing team events. He came up with the idea. And what's great about it all? Steve Kerr listened and incorporated his idea. That's the key piece about being futuristic. Empower your teams to develop ideas that can change the game and set new ground marks for your ongoing winning success.

When you look at job descriptions for many management roles, you will find a very robust level of background/ qualifications in search of a highly qualified candidate. Examples would include effective communication, analysis of data, coaching and development, problem solving, directing, planning, managing complaints, conflict resolution, and financial forecast. The list of duties for a manager can be quite extensive which says a lot about the responsibility for those in leadership roles. What I have not seen too often is innovation being a part of that role/responsibility list as a leader. And this is where I believe leadership has really evolved due to today's technology. Become futuristic!

With so much information available to you through the

internet, applications, social media, and the news, innovation is happening and being reported each and every day. Your mind should literally go through a shift in naturally thinking about an innovative strategy for your business and teams.

I'm a faithful follower of Linkedin and I enjoy my McKinsey Insights application on my phone. I follow companies such as Amazon, Forbes, Microsoft, and United Health Group. Forbes just released a Linkedin update Wednesday, September 5th, 2018 on the world's top innovative leaders with Jeff Bezos leading the way at number one. We can agree that Mr. Bezos has certainly transcended the consumer shopping experience. My McKinsey application keeps me updated with a thrilling update on how companies are using artificial intelligence. The point here is that innovative thinking is now becoming a priority for leadership.

Being Futuristic is all around us

I'm a big fan of hip hop music. Growing up as a kid, I looked up to my friend Geoff who was like a big brother to me. Geoff was four years older than I and he would take me with him to play ball since I was 13. What was even cooler was Geoff could DJ. He had the turntables in his room and he would invite me over often and I would just listen to him DJ records, scratching and mixing. Geoff had skills. Unfortunately, as the young buck, I had to carry his crate of records when he and his boys would DJ house parties. Dang those records were heavy. Geoff, I still owe you a game of 1-on-1. Thanks for being a big brother!

Fast forward 30 years later. I was visiting BJ, my brother-in-law, in early 2017, in his music studio, and I was amazed at how the future of producing music had changed. No turntable mixing. BJ was showing me how he was able to come up with various musical beats and melodies through the computer where his vinyl records were fed into his laptop. He could do the same with his keyboard and any instrument he chose to feed into his laptop. Just like that, he could produce a sound that would make your head start moving. When finished, he'd download his music to a thumb drive and he would be on his way. To me this was straight futuristic work. Shout out to BJ and his FOE Family Over Everything music business. He makes that really good feel-good music. And he's the producer of the Mike Method introduction beat. Good work brother-in-law!

Here's the cool thing. You don't have to be the most innovative person to come up with all this creative stuff. Remember, we are leaders of people and not necessarily the process. This is where you leverage the strength of your team to help come up with the next greatest innovative strategy or process change. I get it. Most companies, churches, sports organizations have teams in place to do a lot of the change work. There is a hunger and need to have someone constantly assessing the market to understand what's changing and how one stays ahead with a winning advantage.

I'll share with you one of my areas of struggle is that I'm not the most innovative person when it comes to initiating the next biggest idea. I can take an issue that has been given to me with their idea and then build based on their innovative thought. But literally in my career, I'd be honest in that coming up with the greatest brand new idea has not been my asset. I'm a people guy and natural problem solver. However, I've understood that I must use the talent within my teams and this is where I've generated the bulk of my improvement wins. Again, this happens everywhere. I guarantee you it's not your head coach coming up with all the great ideas and playbook strategies. His staff is feeding him some very good recommendations by working closely with the players and their very own film review.

So today as I'm mentoring leaders in many capacities, I'm always challenging their thinking through what I call, "Thing Big" strategy sessions. This is no different from what many companies have today, but maybe a different naming convention. Just in my opinion the words "Think Big" prompt me directly to the purpose of the session. Not to just come up with any idea for the sake of it, but a BIG idea that is futuristic and game-changing. I want teams to challenge the status quo. Again, the good news is that the ideas are not necessarily expected to be yours.

Think back to principle 2, Find the Talent in you. Each member of your team, especially if recruited effectively, should have a wide range of talents. Allow those talents to exchange dialogue-building energy within their brains to come up with the next improvement for your company or organization.

Here's the tricky part. We know those placed in leadership roles have a lot of responsibility on their plate. Earlier in this chapter I called out many of the duties and responsibilities in management roles. So, your days are going to be moving at lightning speed, trying to manage your pace and time. As leaders, you have to build a routine of slowing down so you can peel back situations to learn and think through future changes.

So you need a team of talent around you who can help generate this new level of thinking in a proactive manner. Again, we want to lead the changes and not simply react to the demands that other changes be implemented.

There is a great exercise I've led on a couple of occasions that I thought was very helpful in ensuring we are stretching our minds.

Exercise:

1. Write your name in cursive with your natural hand.
2. And then you try writing your name in cursive with your non-writing hand.

Unless you are that darn good, you should have noticed the following: 1. You wrote your name at a slower pace. 2. More of a critical thinking strategy in how you would write in a less comfortable position. 3. You may have tried writing your name a second time because you see the possibility of succeeding at something that may have been unthinkable. The point is that you need to take a moment and slow down and think through improvements.

If I was recruiting a leader to help run my business or an office team, I'm looking for one who has a futuristic thinking. Yes, while executing the day to day priorities are important, I'm certainly taking that leader who can help transform my team with innovation. More specifically, it is a leader who can tap into the innovative minds of their teams!

TMM Time to Ignite "Futuristic" thinking

- Schedule a monthly 1-2 hour "Think Big" strategy session. Make sure the meeting is scheduled on calendars so your teams feel committed to the thinking.
- Create a reward system for recognizing those who come up with the "Think Big" ideas that get implemented
- Utilize social media such as Linkedin and Twitter to connect with highlights of innovation across the globe
- Bring in an outside coach or consultant to help lead your teams through an annual "Think Big Strategy" session. This should be at minimum a full day off-site experience with specific goals, case studies and expected results
- SLOW DOWN. If you find yourself constantly saying, "Yes, I'll get it done," then incorporate a "Why" recap at the end of your discussions. Yes, a "Why" recap. This allows you to pump the brakes and really think about your solution. It may need to be bigger than what you originally had in mind.

Check In point #2

Let's take a break and reflect for a bit. Within the first 4 chapters we focused mainly on the foundational build-up of leading teams. These were the core principles that need to happen before you dive into the day to day engagement and your culture of leadership. I consider the first four chapters similar to the basic framework of building a house. You prepare your construction site working on the framing, plumbing, electrical, insulation etc then you jump into the other facets of the home building project.

These last four principles we just reviewed took us more into the mindset of effective leadership. Similar to building a home, principles 4-7 represent that next phase of your home building project. Since the foundation is now set, how do we begin to take your teams to the next level of performance. We talked about being **flexible**, **failing**, **facilitating**, and having a **futuristic** way of thinking. Essentially we got into the core of leading a change-driven team. And what it takes as leaders to deliver change in an appropriate manner where your teams trust the direction of your vision and plan. This piece is not easy because here we as leaders have to let go of our ego a bit and do things differently to inspire your team.

The good news is that you have gotten through the toughest principles in my opinion. Leading change ain't easy. It takes resilience and an open mind to think differently. If you have this piece down pat, the frustrations of leading will certainly be fixed. Wait, is Fix another F word? Ha,ha

Similar to the College Basketball NCAA March Madness, Let's go to the final four principles and bring this home!

TMM PRINCIPLE #9:

FOCUS – USE YOUR GPS TO GET TO YOUR DESTINATION

One of the more interesting debates you can have is with someone who has an iphone. I was a Samsung Galaxy cell phone user for years. My wife, kids, brother, and many of my friends each have the iphone. My brother, Dexter, was a converted iphone owner leaving the galaxy world. He was visiting with me recently from Virginia and he was showing off his new iphone. He brought up the subject: "Why do you still have the Galaxy?" I laughed because we both agreed Galaxy had the better navigation feature. For me, GPS and email was really what I needed on my phone. He laughed and told me to get with the iphone program. He also reminded me iphone has Siri and the same google maps so I had to get with the program. Those daggone iphone users! Oh well. I finally gave up my Galaxy. Plus, my daughter made me switch so we can facetime. Sheesh. Smile, princess!

What's up with the GPS!

The GPS technology I believe is one of the best examples of being focused and having a plan. Think about it. You enter the destination and hit start. The GPS will quickly map out your destination, giving you the number of miles and estimated arrival time. What's even cooler is that the GPS these days will recognize traffic jams or any construction happening and will detour your route so you reach your destination in a timely manner. The goal of the GPS is to keep you on an intentional plan while removing obstacles. What I like most is that the GPS, even through the obstacles, will never advise you to make a u-turn and go back home. I'll repeat this statement. The GPS will never advise you to make a u-turn and return home. How many of us have a destination where obstacles get in the way, and we stop. We revert back to our normal comfortable way of living and put our goals and dreams on hold. That's been me two, three, four times. I know God may be shaking his head at me saying...."Dude, can you not stay focused?"

Lesson learned – Starting the day with no GPS!

Off to another workweek as the alarm sounds. It's a Monday morning and we all go through this at least once. You wake up, fix a cup of coffee and start checking emails, critical business or news media sites to get an early start to the day. You can thank the cell phone creators for making this so convenient but distracting for us. You should be exercising or perhaps using the time to work on your own self-development. But hey. That's what we do!

Here's what happens!

It is 3:30 am one morning and I woke up thinking futuristically regarding an innovative way to implement some changes for a team within a company for whom I was consulting. I then started thinking about the tremendous amount of work that needed to be done over the next six months to satisfy my contract. I was in over my head. In addition the honey to-do list that I was behind on. You know how it is sometimes. My wife says, "Remember to take the clothes to the cleaners." "This is your week to pick up Taylor." "Don't forget we need to pay for Jordan's AAU registration...etc." It was everything hitting me at once.

So I don't lose my thoughts, I reached over to grab my cell phone and start entering notes while in bed. Of course my wife, like many wives, have this intuition and she popped up yelling that I've disrupted her sleep. Unfortunately, I was pushed to the couch. Now I'm in the dog house for waking up my wife. Fellas, before I even finish this chapter let me state... If you decide to do emails on your phone, just go ahead and get up and take it in the other room. My wife ain't the one for breaking her sleep and most likely yours is the same!

I was then thinking about my 4-year-old daughter at the time, who can be a little cranky in the morning. She can be a handful when she wakes up. It was my fault allowing her to stay up late to watch her TV show. Shame on me. But she's daddy's little girl so I can't be too mad. This was going to be a busy day of chaos!

After dosing back to sleep, I woke up at 5:00 am. I

started working on emails so I could get caught up on a few important administrative pieces before I dove into this critical project. As I made my way through my emails, I was failing to mark what was important for me to follow up on. I believed I had it in my head. Normally, I have a top two list of those key critical items I must complete. But my focus was just to clean out my email box. Keep in mind I had other companies I was contracting for so it was a lot.

It was now about 6:15 am. Making sure Jordan is up. Helping my wife get things ready for Taylor. We were behind because she struggled waking up. MY FAULT! And we were racing out the door at 6:45 am as Dad has to drop Taylor off at school. Shucks.

Now, I'm en route to the office as I'm sitting through traffic. I'm thinking about this project, listening to NBA radio, then switching to some hip hop music to get me hyped as I pulled into the office parking lot. But here's the deal. I reached the office, and the company I was contracting for wanted to shift their improvement focus which required technological improvements. This was not my strong suit. I was contracted to do process improvement, but exercise a greater focus on leadership development. Technology is something I normally do not do. In short, I was spending time trying to find a technology consultant to hire ASAP. Remember those top two items I mentioned I had as a priority. They totally slipped my mind considering the meetings scheduled throughout the day. I wasn't sure how to adjust. I started cancelling meetings yet this was going to put me behind for the week.

Plus I forgot Monday was my day to leave at 4:30 to

pick up Taylor. I promised my wife I would switch up so she could work late. Dammit! At this point it was nothing but chaos but more importantly, I didn't have a plan to fall back on to allow me to make effective audibles to ensure I maintain and get things done. I was just showing up with my work schedule for the day to *manage me*. A lot did not get done that day. I wasn't Focused!

Let's talk about Focus

Focus is a word we hear quite often. As a child, you may have been told on many occasions. Focus on your homework. Focus on your studying. Focus on the chores. Focus at church. Focus on the ball when up to bat. Focus on the rim when at the free throw line. Focus on your routine before you begin your dance. Focus on getting focused. Funny now as I'm a parent, I'm saying the same thing. "Taylor, you don't focus!" I don't tell her that anymore because now she tells me the same thing. Smile, princess! 😊

Interestingly enough, I never recalled my mother or father telling me to be focused about playing outside or playing video games. Which means we are in control of what is priority and where our attention should be placed. The challenge is how to remain focused when there is so much activity around you, especially in leadership roles or as business owners.

I was at lunch one day with a group of professionals and the question was asked, "How in the world can you stay focused with so many damn priorities coming at you?" We just all began venting about home responsibilities with our spouse and kids. And then, of course, work could send you on a wild Field Day race at times. Man, if you are a successful entrepreneur like my guy, Tim, his day is even crazier. Remember I failed at my entrepreneurship but I remember what those days were like per the example I just shared. It's a different struggle for sure. Much love, Tim, for keeping your business going even through the tough times. Your resilience is one I greatly appreciate and admire!

We exchanged stories of the week where we faced challenges that came up as unplanned and how it could throw us off rhythm. Either you fell behind, forgot about the problem, or more priorities were loaded on your plate and you sporadically kept trying to get things done. Where I failed in my entrepreneurship and life in general was largely not having an intentional PLAN and AUDIBLE.

Back to the GPS Feature

When I have speaking events or am doing 1-on-1 coaching especially about life and leadership, I use the GPS analogy all the time. Honestly, this is my favorite story line on how to focus and the importance of having an audible.

We all know the GPS is used to get us to our destination. We love it. It conveniently helps us get to where we need to be. Before we leave the house, we plug in our address and without a doubt we trust the directions the GPS populates and feel confident it will get us to our destination.

If you notice, your GPS never loses focus and goes off track of the destination. It is focused. So focused, that even with a traffic jam or construction, you may receive an alert by which your GPS will reroute you to ensure you reach your destination. It may take you a bit longer but you arrive. The GPS has built-in AUDIBLE programming to ensure it remains focused on its destination.

So what if we had our own built-in audible to ensure we stay on task. My days of getting sidetracked as in my example at the beginning of the chapter would have certainly gotten me to my destination. Stay focused!

One of my good friends gave a cool analogy of sports. He said it's like a quarterback who starts off hot the first quarter completing his passes and scoring touchdowns. The defense may have been in a simple defensive scheme that allowed easy passes to be made. But then the next quarter the opposing team's defense began switching up their scheme with more rushes and blitz packages applying pressure to the quarterback. The coach did not help the quarterback become prepared and the guy began throwing incompletions and interceptions. The quarterback did not have a plan to handle complex defenses so the confidence and focus was lost and the team lost. No GPS Plan!

My first childhood Focus Moment – Mr. Robert was focused!

When I was age four back in 1979, I was fishing with my dad and Mr. Robert in Roanoke, VA. This would be an amusing experience. Mr. Robert was like a grandfather to me. We would fish at a local lake in Roanoke and he had an interesting approach to fishing. He would post his rod on a stake in the ground, while he sat in his chair with a can of beer and legs crossed. He never reacted to a slight nibble on the rod. He literally would sit calm in his chair and continue to sip on his Schlitz malt liquor bull. Meanwhile if my dad and I had a nibble or two, we were jerking that rod back pulling some fish in. But Mr. Robert reacted only if the rod had a strong jerk. He'd rise quickly and grab his rod, reeling in a sizeable trout or catfish. It was amazing. He literally would not react or deviate from his strategy, sticking to his plan. My dad and I would pick up all these smaller fish that were almost too small to eat so we were spending more time working and re-working, throwing fish back into the water. Mr. Robert would leave the day with maybe six or seven large sized fish with minimal effort. My dad and I would have all these small to mid-sized fish in our cooler, exhausted from the work we put in. I remember asking, "Mr. Robert, how do you get all the big fish?"

He would laugh and tell me, "Little Ricky, I have a plan and I remain focused!" Funny thing is his plan did backfire once. One day that light nibble turned into an aggressive jerk of the rod. And the fish pulled the fishing rod off the stake and into the water. Mr. Robert popped up with beer spilling on his pants and his cigarette flying as he chased the rod. Still to this day that makes me laugh! Love you, Mr. Robert!

So it clicked. I did not have a plan for me at work on how I would execute for the day. Those days of frustration were self-inflicted. I was blaming everyone else and did not gain control. I would get it done but it often led to frustration and tired days. I had no plan…no audible so the day truly managed me.

Go back to the beginning of the chapter where I talked about waking up early and working on emails. Those were emails that I allowed to linger from the week prior. I did not have a plan to ensure those emails were up to date daily. I had time blocked on my calendar to review; however, I would dismiss my calendar alert and keep working on something else. I had no GPS function to ensure I reached my destination. I created my own chaos! My point is this. I should have had a routine where I was at minimum planning my day generally despite potential obstacles in order to avoid the chaos.

This Manager Always kept a good GPS Signal

Meet my friend, Keith. Keith was a manager in a company out of Dallas, TX. Wait. Go Cowboys!

Keith was never stressed. When we had our happy hour vent discussions about work, he'd always be sitting back all chill and just listening. Keith was never stressed about work. While me and my other colleagues vented about work, he'd be smiling, eating his wings, and sippin his beer. Keith was a high performing manager and was consistently at the top of his incentive rankings. He was very efficient with great attention to detail in the work he produced.

So I asked Keith later that day, "What's the secret to your consistency? You are the only one that never vents and you keep a calmness about you."

Keith said that he has a plan for the week and the current day and he does his best to stick to it.

- Arrived to work 40 minutes early every day
- Completed a quick review of emails that he missed the day prior
- Created his PLAN for the day, often including accounting review, sales reporting, and company news.
- Kept 30 minutes blocked for lunch on his calendar.
- Accepted no meetings during the last hour of his shift unless critical. Instead, he caught up on important emails or other tasks.
- Not afraid to say "no" to tasks that expanded his workload.

I said to Keith, "Damn, do you stay true to this routine?" He said yes, otherwise the day would manage him and he would be like the other managers in his office, working 10-11 hours per day, stressed out and FRUSTRATED going home. He felt it was important to have balance for his family. Wow! I thought.

He was correct. He truly had a game plan. Keith was focused and intentional about his days. He said his people were always the No. 1 priority. In contrast, many managers are looking for fame by taking on too much, hoping for exposure and the next big promotion. If you are not careful, taking on too much can backfire. I'll raise my hand as I have sat in that seat before.

In some situations, I should have done a better job mirroring some of Keith's practices. Unfortunately, I struggled with focus. Plus I talked a lot.

It took me a few years later in my career to truly get used to developing some of Keith's discipline for myself.

I really had to do this when I started my own business. Time was valuable and I had no one to delegate to. I came up with a plan to manage my time, be organized, and stay focused. This has truly changed my life both professionally and personally, though I'm not always perfect at it.

Richard's Daily GPS Focus Routine

- Do not wake up or go to bed with your cell phone unless all social media applications are turned off. Listen, if you are going to be watching Facebook and IG updates and waking up first thing doing the same you are not focused. You and I both know there are way too many distractions on social media. No way you can be prepped and ready to maximize your day with social media being top of mind. Read a book, or invest into some type of learning app. Just something that is targeting your development as a person and professional.
- 30 minutes daily of "I AM" (It's About Me) Ready Time. This is time away from any distraction. I'll start with prayer, read a Bible verse, and review a leadership quote, and stretch for maybe five minutes. I found stretching helps me to think and remind myself about being FLEXIBLE for change
- 20 minute daily review of my calendar for a 30-day outlook of what's on my plate. This was important. I've made mistakes in which I've scheduled conflicting meetings or personal appointments. I would blame my wife for not telling me about a doctor's appt or

our daughter's parent/teacher meeting. However, our kid's school calendar was right in front me on the refrigerator door. That's a tough argument to win!

- Daily at noon I'd get a Bible verse task pop-up alert. At the mid-day point of the day, chaos typically kicks in and I can get off focus. This is a 5-minute brief getaway mentally and spiritually to help reset my day. I'll read the quick verse or say a prayer to get that FOCUS and spark going.
- Workout four days a week. Remember stay "Fit" to lead. With my son 13 at this time, he was my added motivation to work out. 1x1 basketball was my favorite so I could remind him who's the true baller.
- 2/3 days a week of self-development. I used this time for either reading and/or completing leadership courses so I kept my skills sharpened as a leader and speaker.
- My personal Mike family accounting review. Every Monday, I would review the bills and evaluate savings plan. After a weekend of possible over-spending, I needed to make sure we stayed FOCUSED on our financial goals. Plus, who wants to deal with paying bills throughout the work week. There's enough going on!
- Sunday evening family prayer. With so much going on during the weeks with us as parents and our kids, I found our Sunday family prayer routine to be extremely important. It was one opportunity for us all spiritually to connect as we approached the week.

- Each evening, I'd spend 15 minutes reviewing my task list and checking off what was completed. This is tough because I loop in both my personal life and work to-dos all in one. It's a lot but I need to have that visual check of what's been done and what is pending for completion. I feel good going to bed not having to wake up in the middle of the night thinking about some random "to do" item that's going to backfire on me in the morning.
- This may be the most important planning requirement. Prep your coffee the evening before or nowadays use a Keurig. There is nothing like coming downstairs and seeing that coffee already in the pot. Smile for my coffee lovers!

You may say some of this does not sound like being focused at work. But it does. These other factors will compound the challenges that you have at work. You don't want to be at work while thinking about bill payments, child responsibilities, etc. Remember, we must have a personal/work life balance. Yes, this takes a lot of work. But you are leaders. Great leaders understand it may take some extra to be the best leaders for your teams

There you have it. These are some basic ways to instill a disciplined routine of FOCUS. Try it. What up, Keith? Thanks. You would be proud of me!

You MUST get you an Accountability Coach!

There is a movie called War Room that was released in 2015. This was a popular film illustrating the importance of fighting for a marriage through prayer. The story shows a young and successful African American family who lived in a suburban town. The husband was a successful businessman and the wife, a successful realtor. Through their success both financially and professionally, they encountered what many marriages experience: the struggle to keep a happy relationship. The wife felt the marriage would eventually come to an end and she was heartbroken.

The wife met an older woman who became a close friend, sharing wisdom, as she was easy to talk to. The woman, Ms. Clara, sensed the wife was unhappy and advised she needed to connect with God and pray specifically for her marriage and her own strength. But the challenge was not to simply pray, but to find an isolated room which was a closet. The closet represented a small space with four walls where the prayer could not be interrupted. The lady, Ms Clara, called this room the War Room. And this room was where the wife was going to battle for her marriage through her prayers with God.

Ms Clara essentially became the wife's accountability coach. For the wife to stay on course, she needed someone in her corner to ensure she followed through on the changes that needed to be made. We will talk about follow through in our last TMM principle.

At the end the husband and wife were able to save their marriage because of the guidance of an accountability coach.

Now why did I use the analogy of marriage in a leadership book? You know, just like me. The demands of being a great leader within your profession can impact your relationship at home. Like Keith, he understood as he almost lost his marriage because of his career. That's why he made the changes.

As great as it is to have a plan and to be intentional, you need an accountability coach. We need someone to help identify blind spots to keep us on course. The accountability coach is the GPS logic that helps identify distractions, offer detours, estimate your destination timing, and the voice that reminds you how far you have to go before your destination is reached. Stay focused!

Time to Ignite your GPS Focus

- Develop an intentional plan that will allow you to be focused on accomplishing your desired goals both personally and professionally. For starters, narrow your list down to two each. Too many on the paper will distract you.
- Schedule your daily "I AM Ready" time and stick to it. Your growth is important. You may have to get up an hour early or stay up an hour later. I know it's tough. But if your job needed you at work earlier or later, you would make the adjustment, right? This is for you. Make the adjustment!
- Create a daily routine specific to planning each day, and prioritizing key areas of focus. Go old school. Use a hard copy desk calendar so it's visible—not your iphone. If you use your phone, this planning time will be overshadowed by Facebook, IG, TikTok, Work emails, Twitter, and all the other apps you have on your phone.

- Create your own success metric – If you complete 80% of your tasks within the week, treat yourself to your favorite "I AM" activity. I enjoy craft beer so I'll reward myself by hanging out at the local brewery on a Saturday or Friday evening. If you like getting manicures/pedicures, shopping, golf, reward yourself. But be honest about it!
- Find an **accountability partner.** This may be the most critical area to ensure you remain focused. You can have a well documented plan, but it can easily be distracted by daily events. I have a few for spiritual, professional, financial, and fatherhood. It truly takes another voice at times to keep you on track. Find One Today!

TMM PRINCIPLE #10:

FIT – ARE YOU FIT TO LEAD?

Here we go. 5:00 am my alarm went off on Monday morning in October 2013. I rose quickly to freshen up and put on my athletic gear to meet my good friend, Mike. I actually call him Big Mike because he's about 6'5 and looks like a NFL football player. Me and Mike went to high school together. He graduated a year ahead of me and I probably hadn't seen Mike in over 15 years. Mike was a gym warrior as I call it. Any person who consistently wakes up at 4:45 am to be at the gym by 5:15 am M-F in my eyes is a gym warrior.

Mike was a principal at one of the local middle schools in Chesapeake, Virginia with a great reputation of helping students. We had a lot in common. We both were husbands, fathers, had influential roles at our place of employment, and were trying to manage the everyday stress like many.

The Friday prior, I ran into Big Mike at the gym. I was not a morning person but felt I needed an early exercise pump. I like to work out but not so much to where I want to wake up early in the morning but it was needed. Both Mike and I had fairly demanding careers so it was good just to catch up and talk about life while simultaneously getting a workout.

I was telling Big Mike about one of my partners who works in a very demanding leadership role for his company and had been venting about his stress level. Coming from Norfolk, Virginia, my hometown, James was doing very well working for a big time company in New York City. Despite the big salary and lifestyle he was stressing himself out at work.

James like me had a high drive for excellence. He wanted to be #1—to be the best at his job. He wanted the best out of his office. What made matters even tougher was that he worked for a boss that was just as high strung for excellence. Basically he was getting his butt pushed to the limit, and he was pushing his office to the limit. So everyone was stressed like crazy. He would call me and say, "Rich, why aren't my people performing like they should?" He said, "It's like they've been performing well; then all the sudden they're making mistakes. Calling out for work, somebody screwed up a 5MM international wire that went to the wrong account. He said literally that day around 4:00 pm he thought he was going to have a heart attack. He said the office started spinning and he was sweating like crazy and stopped and prayed.

I was like, "Dude….are you serious?"

James nodded vigorously. "Rich, I'm serious."

Luckily he simply felt lightheaded and that was it. No heart attack.

I told James. "Dude, you're too high intensity. Like, I couldn't work for you, though I love your passion." I said, "Man, your people started messing up because they're burned the heck out."

We talked back and forth and agreed that he needed to do things differently. About two months went by and I called him to check on things. Dude was doing outstanding. He said he was still high demanding but his energy level was up. He found out by going to see his doctor: his blood pressure and sugar levels were high, he was not eating healthy, too much drinking and no exercise. So basically that's why he was so high strung so much. He said he was doing well and exercising and eating right. Good for my guy, James!

I thought to myself. Oh crap I'm just as passionate as he is so I need to make sure I get my butt in shape so I can handle my career and family demands.

Time to get fit!

I arrived at the gym at 5:32 and worked out for an hour. As I was leaving, I ran into Big Mike. Here's where we exchanged phone numbers. He shared with me that he worked out daily and he would be there the following Monday morning if we wanted to work out together.

I got home about 6:40 am and I was hyper as could be. My wife was downstairs prepping the lunches and I was in full energy mode, chasing my 3-year-old daughter around the kitchen. That morning I was truly feeling like I was "fit" to lead!

That Friday at work was one of my more productive

and stress-free days. I had a calendar of meetings and some issues lingering to be fixed. However, I felt in control of my day. I had great energy and more importantly my positivity was greatly noticeable. Around noon, one of my associates said, "Richard, nice to have you back!"

I said, "What do you mean?"

She said, "You weren't the same this week. Me and a couple of other teammates were talking about how they noticed the change."

My morning hellos had been extremely brief and I was not as personable as I normally was. She said, "Richard, it's so crazy but what keeps me going is your laughter, energy, and the fact that you care about us."

Wow, that statement hit home for me. I met Mike that following Monday morning to begin my daily workout routine.

I would begin meeting Mike every morning around 5:20 am, starting that following Monday. My energy grew along with confidence. My kids noticed the change. They'd previously been used to me yelling their names with excitement when I came home to greet them both. For the many prior weeks of stress and not managing my days well, I had no excitement in my voice. But now Mr. Dad, Husband, and Manager was back.

As I started my workout routine with Mike, I cut out fried food and social martinis for 90 days. My body needed to be physically and mentally fit to combat the stress. I could feel, the harder I pushed in the gym, the sharper my mind became at work.

For the next 90 days my body went through a

transformation in shaving off about eight pounds of fat while gaining solid muscle. I was nowhere the size of big Mike, but I felt fairly good looking in the mirror. My wife would even tell me both my stomach and buttocks went down. When you get compliments from your wife that you are slimming down and looking good, you are WINNING!

More importantly, the transformation began to occur at work. I began to have more control of my days. Versus reacting to issues, I felt more alert, asking more questions about problems that needed to be fixed. I became a better communicator with my teams. I was more in tune to their workload, moving work around, and pulling in extra help. I had to go back to the FUNDAMENTALS. (Remember Fundamentals in chapter 3).

Much of the ownership started with me as their leader. I had to become more fit to keep up with the change and chaos of the business. When I'm talking to other leaders at networking events, I enjoy asking this "out of the box" question. When they talk about their stress and their team's struggle I can obviously relate. So I'll ask, "How physically and mentally fit are you and your team?" Most are stunned by the question. Many will think twice before attempting to answer.

I respect one gentleman named Matt whom I met at the event. He honestly shared he was not in good shape. He then asked me, "What does that have to do with my role as a manager?"

I just shared based on my experience, admitting that I was not a health expert, but the energy level that you obtain from working out translates to the office. Industry studies show that a health and wellness plan helped to stimulate brain cell growth. In short, your memory and learning capabilities increase. We exchanged business cards but I did not expect to hear from him.

Four months later, Matt sent me an email expressing his thanks. He shared that his company brought in a health coach to do seminars on health and wellness and how he and members of his team had weekly health plans. He acknowledged his increased energy and that his fellow office workers recognized hisreduced stress levels. That was awesome!

Use the Sports Team approach when it comes to Nutrition

A very good article I came across written in 2013 regarding Chip Kelly, former Oregon College football coach. The article captured the culture of nutrition Chip Kelly and his assistant, James Harris, had established for the team. Chip Kelly was the head coach of Oregon football from 2007-2012. The Oregon football team was known as a speed team, called Tracktown because they had so many players with elite speed among different positions. Chip Kelly and his team were able to achieve some impressive victories in his tenure, winning the 2012 Rose Bowl. As people remember Chip Kelly for having a speed team in college, many may not be aware of the focus he had on his players being fit through nutritional science.

Chip Kelly was often criticized for having such a perceived insane approach to nutrition. But, he was able to convince the Oregon school to fund nearly 1 MM on the nutritional culture. With the help of nutritional experts, he was able to build a business case that captured the advantages of having individual health plans based on each player's physical make up and position.

After reading about Chip Kelly's belief system on nutrition, I realized a change needed to be made when it comes to leadership. If, as leaders, you want to have the right influence and stamina for long and challenging days, there needs to be more promotion of a healthy lifestyle. I've been to a number of health and wellness seminars and the facts to me appear as an easy sell. If being fit helps to reduce heart attacks, diabetes, stress, and other ailments of the body and mind, why not jump on board? But it begins with you as their leader!

I'm telling you it has made a difference in my life overall. And truly, when I fall off I notice a considerable difference. For those of you who do not like the gym scene and don't want to be around all the muscle bound superstars, no problem. Here are a few recommendations. I'm not a health and fitness mastermind—only sharing what's worked for me.

1. Start off walking in your neighborhood or local park.
2. Purchase a treadmill for your home. Buy an affordable weight set and put it in your garage or man cave
3. Cut out the sweets and the sodas to a minimum during the week. If you make it through the week, then maybe the banana split ice cream on Saturday is warranted. 😊
4. Eat healthy during the week. I know it's literally

impossible these days to cook dinner every day at home, especially when you are on the go. I try to enforce a one weekday eat out day. We call it Chick-Fil-A Wednesdays. This started during the Pandemic and became our eat-out day. But honestly with Jordan's basketball and Taylor "Girls on the Run," we do cheat.
5. Pack your lunch for work. Listen, if you don't, you will be eating at a fast food spot every day and it's going to hurt you. Eat at work and then maybe take a walk. If you do eat out, consider a salad. I'll give credit to my former boss; he was good at eating salads daily.
6. Once you build confidence, get your gym membership. Honestly even if you don't like the gym, it's a good getaway. Hook up your Air Pods and zone out. Hop on the treadmill, bicycle, and take off, watching ESPN on the TV screen.

Again, I'm not a huge health nut but I understand taking some type of intentional action is good for your mind and body. By the way, I've lost 15 lbs. so it must work!

Time to Ignite a "Fit" and Healthy Routine

- Partner with other groups to help create a more healthy and work-life balance for your teams. Forbes indicated in an October 2017 article, that nearly two thirds of employers offered some type of wellness programs.
- Become extremely knowledgeable of the health and

wellness plans that are available to your teams.
- Organize a "Get Healthy" Campaign – My Linkedin news updates frequently shows a network of professionals taking action to get fit. The pictures are amazing and it looks "Fun" (Remember our F principle Fun)
- Include health tips of the week within your huddle agendas to keep health and wellness visible among employees.
- Find out who is very health-conscious when it comes to working out and nutrition. Encourage them to be your health champion and they will run with it. I've seen the health champion who had the teams out walking on lunch break and after work, and you only saw healthy food at their desks. Great job!
- If your workplace allows, organize an on-site yoga class to help with the stress release during the day. This is becoming a trend within the office space.
- Incorporate quiet rooms on your floor. Convert old offices or conference rooms into quiet rooms. These can be used for brief naps and stress getaways.
- Lead from the front – As a leader you have to show why being healthy is beneficial. Your employees need to see how being fit has benefited you. Share how exercising daily and a good diet has helped manage your stress levels.

TMM PRINCIPLE #11:

FEARLESS – TURN YOUR HAT BACKWARDS AND JUMP IN!

People always ask me why I wear my cap backwards. Sometimes it's style and fashion. But more likely it's because this bald head of mine is too big to wear my hat forward like I'm supposed to. LOL.

Seriously there is a reason why I wear my baseball caps backwards. If you watch baseball, there is one player on the field who turns their hat backwards when they hit the field. And he or she is likely the one who has the dirtiest uniform throughout the game. It is the catcher. Yeah, I get it. They wear a protective helmet so wearing the hat backwards is likely their only option. But the catcher is the one who at all times is in the midst of the game. And you may say they take a lot of risk. A bad throw and they are diving left to right to stop the pitch from getting behind them. If someone is attempting to run home for a score, moving full speed, your catcher is the one trying to tag him out, often getting run over. And heck, every time the batter swings, it just seems risky that the catcher may get beaned by the bat.

Catchers are on the ground every play of the game for their team. They are not afraid to get dirty. Catchers are fearless, understanding their presence can have a vital impact on the game. Catchers are often the calming voice to the pitcher. When the pitcher has bad throws before the coach approaches the mound, it's typically the catcher who talks to the pitcher and explains how to improve the pitch. The catcher helps prevent the pitcher from becoming frustrated. Catchers wear their caps backwards!

So often in leadership you have to be ready to dig in the trenches and be fearless, leading your teams through critical changes and the chaos that comes with it. The adversity can be frustrating. The long hours and challenges to overcome can be painful and disturbing. This is when you have to be hands-on, leading from the front, demonstrating leadership that is resilient and respected by your teams.

The Movie Gladiator

One of my all time favorite movies is Gladiator with Russel Crowe who played the star character by the name of Maximus. The movie was released in 2000. I've always been a fan of Roman gladiator movies growing up. It was the chariot, horses, and armored gear with swords that seemed exciting. Maximus was a proven general who served the Roman Emperor. In this movie Maximus was like a son to the Emperor, though the emperor had his own biological son named Commodus. Commodus later killed his father because the emperor chose Maximus to retain his throne upon his death over his son. Out of rage Commodus killed his father. Later he had Maximus' wife and son killed, though Maximus survived.

Barely surviving his planned death, Maximus was found and picked up to be a slave to join a team of gladiators. If you watched the movie, Maximus was able to gain revenge by killing the son at the end of the movie.

Maximus was a great example of a fearless leader. He led a team of other captured men by verbal silence but inspiring action. These men saw his ability to lead as they conquered several gladiator competitions where there really was little chance for this team to survive and win. There was a point in the movie in which Maximus could have prematurely reacted to a chance to kill the king but he realized there needed to be a plan. Maximus always got dirty in his competition leading from the front. He didn't require yelling and 'in your face' motivation to inspire his soldiers. He led by action and they knew he was fearless.

I use this example because I believe your leadership today has to be fearless. Fearlessness does not necessarily mean going around battling the world and ramming through every decision for your respective business or teams. That's not a fearless leader. Fearless leaders do not react when under pressure. They recognize there is a plan that needs to be executed. Even during moments of sudden obstacles and mistakes, fearless leaders remain confident and tactful in how they execute the daily efforts of their business and teams.

Somebody on the team made a HUGE mistake!

You've had a long exhausting day of meetings and deadlines that required much of your attention throughout the day. Not only that, your trusted assistant was out for the day which disrupted your plans. Your boss needs you to finalize the quarterly budget numbers by end of day. You have one of your more high-profile clients who wants to expand their advertising and marketing footprint and they want a proposal by today. Another competitor has jumped in and caught their attention. This is something you would normally give to Janet your top marketing analyst but she is out on vacation You review your TMM talent matrix and assign the proposal to be completed by Joe. Joe is familiar with these types of proposals from helping Janet in the past. Plus, Joe has a lot of creativity. Joe completes the proposal and sends the details to your client. Joe calls you with excitement that the task is complete and is confident the client will be pleased.

You leave for the day and receive a phone call from the client yelling to the top of their lungs. They are yelling about the projected costs being severely high and the lack of consideration for their long standing relationship. You make calls to the office and you find out that Joe had some incorrect formulas in his calculation causing the #'s to be off.

When you get home you were able to look at the report and talk Joe through the changes. Client calls back wanting an immediate answer regarding your proposal. You explained the error and send back the corrected proposal.

False Alarm! The client was still livid and was threatening to withdraw their business from your company. Talk about FRUSTRATION build-up! You trusted Joe and he did not deliver. Maybe yelling at Joe and potentially terminating him for the error could be understandable. But a fearless leader takes full ownership handles the situation with confidence knowing this an opportunity for a teaching moment. Joe feels his job is on the line when you sit down with him. But you coach Joe through the process:

1. Was there anything about the request that may have been unclear?
2. Were there distractions that may have caused you to mis-key the account number?
3. Did I create an added sense of pressure that may have caused the oversight?
4. What would you do differently to ensure you will not repeat the same error?

You finish the conversation by delivering appropriate coaching. "I understand mistakes happen." "It's also important to understand how critical accuracy is to our business. Moving forward, as you are finalizing this type of proposal request, have one of your teammates perform a second review." I too have made similar mistakes and realize the importance of having another set of eyes to help verify your work."

Joe already feels bad about the error. But now you just ignited a level of confidence through your coaching that represented a sense of care and support for Joe. Joe is not likely to make this mistake again. He's also thankful for the trust his manager showed in him, not pointing the finger but coaching him through the process. Trust me these types of scenarios will happen. Your fearless approach will be crucial on how you handle these situations.

One who only "manages" and does not understand the value of fearless leadership would have demonstrated a totally opposite reaction. The manager may have yelled, finger pointed, demoted, possibly terminated him, but more importantly would have placed all the blame on Joe. This particular manager is not going to have a lot of people working for him or her. Or they will have an unhappy team of employees who won't bring their full potential to the office because they fear for their jobs every day.

I share this example because I understand the pressures that we go through as leaders every day. Whether you are running your own business or leading a team, there are going to be roadblocks that disrupt your day. It may not be your team. It may be your boss who at the last minute pulls you into a "fire drill" that requires an urgent fix for which you were not prepared.

A fearless leader cannot become down, side-tracked, intimidated, rattled, become reactive, or lose confidence in solving the issue while meeting the needs of your team. Why? Your team is watching you. They are watching your reactions, gestures, voice inflection, and choice of words. This would be one of those situations when you are FRUSTRATED, but you quickly say with conviction, "I'm fearless!" You gather yourself and proceed with the actions needed to execute the needs of your team.

Retired NBA Coach Phil Jackson is a Fearless Leader!

Phil Jackson is a retired head coach of both the Chicago Bulls with twin "three-peats"(1991-93, 1996-98) with the Chicago Bulls, and the Los Angeles Lakers. With the Lakers he won five championship titles with a total of 11, coaching three mega superstars, Michael Jordan, Kobe Bryant, and Shaquil O'Neal. Coach Phil Jackson has one hell of a resume.

What used to get under my skin when watching Phil during games was what I perceived to be his nonchalant posture on the sideline. It would frustrate the heck out of me at times watching teams make a run on his team in the 3rd or 4th quarter and he would just sit there relaxed. He wouldn't call timeout to break up the opposing team's pace and momentum. I didn't see him yelling at anybody. It just seemed like "whatever!" I would be watching the games, especially the Lakers games yelling at Phil on my TV and not the players. I was in my 30s during this time. I managed professional work teams as well as coaching little league sports. "Yes, I'm comparing my little league sports coaching to Phil Jackson!" Got to laugh. But my thought was, "Where was the fire and the demand for greatness!" I never could just sit there while my team was getting our butts kicked or playing poorly. I would "yell" at my kids, demanding attention to detail and bringing a heartfelt effort. I'm a huge competitor and I motivate well through high energy and at times a demonstrative behavior. But, only with the kids I may have been a bit over the top! LOL

I read an interview one day that opened my eyes about

Phil Jackson. In this interview, Phil talked about why he does not call timeouts during the times his team is not playing well. Phil actually believed that calling timeout did not show his trust as a coach for his team to manage through the chaos and down times. The leadership growth in his team was the players collectively figuring out how to overcome adversity in heated moments. And during timeouts, he may just pull one player to the side to coach directly, encouraging him to regroup and find the inner strength to build up his teammates.

Because of this routine, his team was able to win so many close games in the final minute because they knew how to handle high intensity moments. They were relaxed and poised, displaying a fearless-minded spirit just as their coach. I thought wow! Phil was really smart and fearless! Most coaches would be yelling in those same intense situations. But Phil was fearless, showing confidence through little words and minimal gestures. Again, he held 11 titles and three of the top 25 NBA players in history. I think we can learn something from Phil. His fearless belief system worked!

Truthfully, my life changed as I was writing this chapter about three years ago–2018. Fearless was a new principle to the Mike Method leadership principles. It was interesting that when I was seeking to become a fired-up leader how both Maximus and Phil Jackson came to my mind as fearless leaders. Listen, there are plenty of fearless leaders out there. Martin Luther King, President Obama, Bill Belichick each of whom you could argue demonstrate a very even-keeled attitude and charisma of leadership that instills confidence.

Change and positive results will occur.

So, I realized I needed to change how I led in my professional career and at home. I'm very intense and want to win. I'm for my people all day but I was one who could get rattled when obstacles hit. I would push teams to the very limit with great love and support. But at times I think my style was intimidating and others did not feel they had a voice. Funny, as I'm writing my father comes to mind. We both inflect an authoritative tone at times. What can I say, we are "Mike men." Smile dad! That's not a good thing when leading teams, as your voice needs to be one that inspires empowerment.

Challenging the status Quo – Can it be done differently?

A fearless leader in my opinion has to be willing to challenge the status quo. Status quo simply means "the existing state." Or I would often say to my teams "simply challenge what we do today. Why does it have to be done this way?" Fearless leaders are not complacent regarding the performance and success of the day. Fearless leaders are thinking about how things can be done better. Here I don't necessarily mean thinking of the greatest technology advance such as the chapter on futuristic advances.

Here's an example of a simple change arising from a fearless challenge of the status quo. Lisa, a good friend of mine, managed a collections call center in Texas in the early 2000s. She and I used to work together back at QVC as young college students. We had a similar leadership style in which we really care for the people paired with a high demand for results. After college, her career took off like crazy. She was managing a team for a couple of contact centers, then got into project management. She was then promoted to help build a collections call center. Lisa was tough when it came to strategic vision and driving results. She really represented leadership—especially as a young African American woman.

She was telling me about a business case she put together to change the hours of her collections center. They were open from 8:00am – 11:00 pm est Monday – Friday and Saturdays, 8:00 am – 6:00 pm est. These were standard hours in most centers she shared. She felt being open past 9:00 pm est yielded very little productivity and it was a waste of hourly wages. Lisa shared her recommendation to close hours at 9:00 pm est by shifting schedules two hours back to beef up staff in what she felt were the primetime hours of the evening 5:00-9:00 pm est.

I remember she called me venting regarding the fact that executives continued to deny her recommendation. But she remained relentless and fearless. She modified her business case, proving the dollar value associated with the change and the productivity gains that could be achieved. It really made sense. She also laid out the fact that there was no technology cost associated, so piloting this idea became a win-win. Over a period of six months, Lisa eventually was able to change the hours in the office while gaining ground-breaking collection dollars each month. What's up, Lisa? You did your thing. Way to stay fearless!

As you are leading through these days of complex

changes and fast-paced demands, remain confident! For me it's like being a husband and a father. Through the roughest days of work, business, finances, you name it, I try to remain calm. If my family sees that I'm worried they will become worried. But if they see the confidence and a smile daily, they will feel secure that even during tough situations I'll be fearless in getting us through any challenge. I'll be frankly honest here. Starting my own business had some sucky moments: Losing contract bids. People not showing up for training sessions. Prospects cancelling appointments at the last minute.

It was frustrating as heck, and there were days I wanted to lower my head when I walked through the front door. And one day my daughter questioned my spirit. She said, "Daddy, you are not yourself today."

I said, "Yes I am, Taylor. She said, "No, daddy, when you pick me up you always talk to me and make me laugh. You are not laughing today." Wow, this was coming from a five-year-old! So, this was my reminder in life to always remain fearless because there cannot be a day when my daughter questions my confidence and attitude. Here I am today writing a book about leadership. "Thank you, Taylor. You reminded daddy to remain FEARLESS!"

TMM Principle #11
Time to Ignite a Fearless Attitude

- Become less reactive. Always attempt to gather as many facts as possible before making any knee jerk decisions
- Fearless leaders are great coaches. Study the art of coaching. There will be issues that compound the needs of your team and mistakes will be made. Be a coach who asks probing questions that show genuine care and support.
- Don't be quiet at the boardroom or virtual meeting room. If your office has been doing the same thing for months or even years, speak up with a recommendation to do things differently. I'd use Simon Sinek's book, *Start with Why*.
- Voice inflection/posture/attitude are all signs that will illustrate your confidence or defeat. Maintain a smile and words of encouragement even through the toughest of challenges. Your teams will go as far as you go.
- "That's my bad!" – Take ownership of mistakes. Though you may be leading your office or team, it's not necessary to appear to be perfect. Fearless leaders make mistakes so no need to perceive yourself as a Mr. or Mrs. Know it All.
- Turn your hat backwards. Remember to get dirty and do the work with your team to reach a solution. Lisa, who I mentioned, definitely wore her hat backwards.

TMM PRINCIPLE #12:

FOLLOW THROUGH – NO EXCUSES. GET IT DONE!

In June 2015, my son came home from a local youth basketball camp at Old Dominion University in Norfolk, Virginia. He loped into the driveway to shoot hoops. He was 10 years old. He made 20 straight free throws.

I said, "Jordan, I didn't realize you could shoot free throws that well. And your shooting form looks AWESOME!" I said, "Dude, why do I have to pay money to send you to a camp to do the things I've been teaching you for years?"

Jordan said, "Well dad, you never taught me the B-E-E-F method for shooting free throws." I asked Jordan what he meant. He stated B-E-E-F represents the following:

B= Balance, Ensure your stance has an appropriate spread of the feet at shoulder width apart
E= Eyes, Your eyes should be focused on the inside of the rim

E= Elbow, Make sure you bend your elbow in creating an L shape release
F= Follow-through, Releasing the shot with complete extension, ensuring the velocity of the ball reaches the trajectory of the basket.

A 10-year-old kid then told me the importance of follow through. If you did not have the appropriate follow through, you will miss the shot. Jordan went on to win the free throw shooting title for his age group at camp. Smart kid!

Follow-Through - Your teams are watching!

Follow-through is defined in the dictionary "as to *complete* a task; to see a task through to its completion." Simply put, you must follow through on the things you start. It's great to have a plan or guide to success; however, your execution is defined and measured by the ability to execute the plan. Truthfully, every F principle to help you avoid the frustrations of leadership will go out the door if we don't FOLLOW THROUGH and take action.

I was reading an article out of the Harvard Business Review regarding the topic: "Can your Employees Really Speak Freely?" The article summarized the fact that employees often do not speak freely because they do not feel their managers follow through on requests and feedback that has been shared. Basically, those in leadership roles will talk the talk but teams do not see the talk being walked. So confidence is lost. It goes back to our first chapter on foundation where I wrote about confidence and trust as a key foundational belief.

When I look back at the Mike Method "F" Principles, they truly represent the plan. TMM uses a simple series of F principles to more easily turn your frustration into actionable leadership techniques. I've expanded on the method over the years from my experience in leading teams and starting my own business.

At the end of the day. Your teams are watching you and holding you to whether you are truly invested in your plan and the people. If you don't follow through, you will immediately lose the trust and confidence of your team. It's no different from parenting and relationships. I quickly lose my daughter's trust when I tell her we're going to get ice cream but then I get tired or sidetracked and end up somehow at the golf course.

"Daddy, I thought we were getting ice cream! This is what you said."

That's a broken-hearted little girl. But if I follow-through she will then have the biggest smile and trust for her dad. This is how it works with your teams. Sorry, princess, Daddy sometimes loses FOCUS! 😊

I truly believe you will be inspired by what you've read. I'm giving you a virtual high five for showing interest in my completed book. There are going to be some principles that stand out more than others. There may be a few where you scratch your head and say 'this don't apply.' You may even say that was a rough chapter to get through. To keep it real, I said the same. LOL

But I believe you will be pumped about sharing what you've learned or ready to implement these principles within your team. You will also find that these principles are relatable to your relationships outside of work. I'll save this for book two! Trust me. I'm working on book two already!

Here's the deal. Many of you have had some really good leadership training or have attended powerful motivational speaking seminars. You left totally inspired and fired up, ready to take on the world. You were so excited to implement this new way of thinking or leadership because of the message you heard. I've been there. You will face something similar once you close my book. It will be no different than when you leave your seminar. You are going to walk right back into change and chaos. You are continuously going to face obstacles, tough areas, and blind spots that are going to disrupt your capacity to lead, and to lead effectively. What I've shared may slowly fade out the backdoor. It's how life goes. It's tough to stay on track so your FOCUS principles have to be fully charged at all times so that your GPS never turns off.

Here's what is important. I haven't read an article to this day that celebrated a famous historian, entrepreneur, celebrity, sports star, or educator, for having a plan. They were celebrated because they followed through in executing their plan. Great leaders such as Martin Luther King Jr, Sean Combs Puff Daddy, Frederick Douglas, Phil Jackson, Serena Williams, Barack Obama, Coach Mike Krzyzewski, Lebron James and Melinda and Bill Gates are known for their execution. They each had a desire to deliver excellence through adopting a plan and endurance to stay the course throughout their journey. This is called Follow Through!

TMM was created more importantly to help give you a

guide to manage the roadblocks of change and chaos that often drive frustration. I understand that leadership comes with great pressure because people are counting on you. Remember within my introduction, I talked about everyone needing your attention—from community, a local sports team, your family, you name it.

I know first-hand regarding the pressures of leading teams and trying simultaneously to be a great parent and spouse. And with so many advances developing through automation, I wanted your solution to be simple and easily relatable. Basically, it comes down to the people. Don't forget about your people. You can now quickly divert your frustration to the solution through the F Principles.

Remember, TMM does not eliminate the obstacles from happening. TMM actually welcomes the distractions because you are going through a growth process of leadership. TMM will now give you the ability to truly create your very own Focus GPS Logic into your leadership that demonstrates a leader who can lead through change and chaos.

As leaders, we are at the free throw line every day with the trust and confidence of our teams and people around us whom we influence. I've equipped you with a game-changing experience that will help grow your leadership within teams. TMM is not about the process but the people we lead.

I've given to you what I've learned through success and failure and what has earned me a creditable reputation for leadership. Also, using my family as examples I've shared a number of situations in my family as learning experiences since many of us have families or eventually one day will.

This will all make sense. This method has helped me to become, not only a proven leader, but also a better father and husband.

So today, take action in making TMM a part of your leadership growth. Lead with enthusiasm and greatness. Balance your presence with your teams. Keep your eyes connected to the needs of your people. Bend your elbow daily giving high five's, recognizing the performance of your team. And most important, complete what you started and Follow Through. You are now ready to make both your professional and personal shots at the free throw line. Thanks, son. You taught me a lot! I love you, dude.

Who says you can't say the "F" Word!

A FEW SHOUT OUT'S!

To my parents Evans and Brenda Mike. I was absolutely blessed to have you both create a culture of leadership growing up as a kid. The FOUNDATION principle began with you. You showed my brother Dexter and I nothing but love, care, work ethic, trust, and responsibility. This has made me a growing leader in both my professional career and as a family man. Love you!

To my wife Toshiba and our kids, AKA "The Mike Team," thank you for the inspiration allowing me to focus in pursuing my purpose. Jordan and Taylor, walk in leadership your own unique way. We are proud of you both. Jordan, continue to FOLLOW THROUGH like you taught me. And for Daddy's princess, stay FOCUSED! 😊

My brother Dexter, you motivated me simply being your older brother. With almost a 12-year age gap, I had to step up my responsibility and lead. I've learned a lot from you and glad we continue to be close today.

George F. Spencer; bro you were the first one to tell me, "Ricky Rude," just ask for help! I didn't get it 4 years ago. I get it now. "Ask for help" has become apart of my leadership journey. Thank you!

A FEW SHOUT OUT'S!

Coach Walt Green, Thanks for seeing my potential as a high school track athlete. My AAU experience traveling to national events changed my life. This inspired me to be FLEXIBLE for change. By the way, this was a time when parents did not travel with the athletes. 😊

My prayer support. Pastor Gary Nelson, Dave, Dino, John and Thomas. Thank you guys for having me in prayer and helping me raise my faith through this entire process.

Finally, a BIG shout out to my friends and extended family. What amazing relationships we have and I truly appreciate you!

ACKNOWLEDGEMENT

To Glenn Proctor, AKA "Coach" my book writing coach. Coach, what can I say, "you kicked my ass" over this book as you would always say. I had to eat some tough criticism from you that frankly pissed me off. I finally learned I had to write this my way coach. We finally got it done. Thanks for pushing and believing in me. Thank you!

Jeff Conley; dude thanks for the unwavering support and taking time to edit my introduction. You have a natural gift for writing. Continue to pursue it as you continue to influence others with your words!

Dr. Kenneth Morton, Thanks for taking the time to help edit my synopsis. This helped position my words for others to quickly understand my vision and purpose for this project.

DEDICATION

I dedicate this book to all the leaders out there who often get frustrated. Dealing with so much frustration taking on leadership within the varying capacities is a heavy weight on your shoulders. I know the expectations are high and I recognize your desire to execute with excellence. I believe in you and I trust your capacity to lead even through all the demands that are in front of you. You Got This!

For my younger readers perhaps coming out of high school or college. I truly believe as you pursue your career aspirations, the principles I've presented will only help you. In a tough market place competing to enroll in the college admissions process, sports and academic scholarships, your first job after college, and perhaps climbing the career ranks. These principles will give you the competitive advantage that sets you apart. Go make a difference and represent our next generation of young leaders. FOCUS & FOLLOW THROUGH!

FRUSTRATION/ FOCUS = FOLLOW THROUGH

The Math Don't Lie!